£7·50

ALUMNI MEMORIAL LIBRARY
Lowell Technological Institute
LOWELL, MASS. 018[illegible]

AF697803

# MAINTENANCE ENGINEERING

## Organisation and Management

# MAINTENANCE ENGINEERING

## Organisation and Management

FRANK GRADON

*C.Eng., F.I.Mech.E.*

143057

APPLIED SCIENCE PUBLISHERS LTD

LONDON

APPLIED SCIENCE PUBLISHERS LTD
RIPPLE ROAD, BARKING, ESSEX, ENGLAND

ISBN: 0 85334 555 4

WITH 105 ILLUSTRATIONS

© APPLIED SCIENCE PUBLISHERS LTD 1973

TS 192 .G7 1973b

All rights reserved. No part of this publication may be reproduced, stored in a retrieval system, or transmitted in any form or by any means, electronic, mechanical, photocopying, recording, or otherwise, without the prior written permission of the publishers, Applied Science Publishers Ltd, Ripple Road, Barking, Essex, England

Printed in Great Britain by Galliard Limited, Great Yarmouth, Norfolk, England.

To

ELIZABETH

*without whose understanding, sacrifice and companionship this book would never have been written*

# CONTENTS

*Chapter* 3

## THE DRAWING OFFICE as an Administrative Element of the Maintenance Organisation



*Chapter* 4

## THE CENTRAL WORKSHOPS Organisation and Management for Maintenance Engineering and Jobbing Work



*Chapter* 5

## INSPECTION AND MAINTENANCE OF SITE PLANT Organisation and Control

*Chapter* 9

## GENERAL ADMINISTRATION with Guides to Efficiency



## APPENDICES

# PREFACE

The increased quantity and variety of plant and machinery employed by industry have led to a recognition that the efficient operation and maintenance of such equipment, much of which is of a highly technical and costly nature, is vital to profitability. The development of the mechanisation and automation of production, with the accompanying development of ancillary services and safety requirements, has emphasised the need for sound maintenance administrative services in all areas of industrial activity. As production has become increasingly mechanised, with an emphasis on full utilisation of the equipment employed and with the ever-growing complexity and size of industrial organisations, the responsibilities of maintenance engineers have proportionately increased and continue to do so.

Today's maintenance engineers must not only possess a sound technical training and practical aptitude; they must also be competent organisers and administrators. This reflects the simple fact that in this highly mechanised era a whole production unit can be brought to a standstill by a minor breakdown in the production chain; when translated into terms of lost production, such events can prove very costly.

Because of the rising costs of labour, substantial savings are being achieved by the continuing mechanisation and automation of production, and the accompanying reduction in man-power demand. On the other hand, the physical processes of maintenance remain essentially manual, and can be satisfactorily sustained only by adequately staffed and efficiently managed maintenance departments. It will thus be very clear that the maintenance function is an integral and essential part of any industrial organisation and plays a major role in achieving overall operating efficiency and profitability. It is also obvious that the effective management and organisation of the maintenance department and its operations is a subject for the most serious study.

The economic potentialities of sound, effective and efficient maintenance management are far-reaching. Because of the costly man-power to be deployed, and the life and utilisation of expensive plant and equipment needing to be maintained at the highest economic level, very large sums are spent annually by industry under this heading. In many cases the sources and nature of these costs are not readily discernible, accountability is difficult to establish, and remedial action is not easy to apply.

The superficial impression of maintenance as a burden to be borne to 'keep the wheels turning', without regard to its true function, is reflected by the lack of attention it has received in many quarters. Today, the sheer force of economics is bringing about a more enlightened attitude and the importance of efficient maintenance administration is increasingly recognised throughout industry.

To be effective, maintenance calls for patient study and painstaking work. Its fundamentals have to be recognised and given clear and concise definition. All of the varieties and complications of modern industry cannot be embraced by a universally applicable scheme but, as with production, the broad principles of management and control can be established, at the same time highlighting the essential differences of the problems to be overcome. The purpose of this exploratory work is to go beyond generalisations and examine, in depth, the various aspects of the subject and the related data and information which can be tailored and adapted by those responsible for maintenance engineering in any industrial organisation.

The author wishes to acknowledge the helpful advice and guidance he has been given by the publishers, and the valuable constructive criticisms offered by their readers.

To the young engineer in the early stages of his career, faced with tasks of seemingly increasing complication, he commends the advice given him, in his student days:

> A complication is only a collection of simple elements knitted together. Break it down into these elements, analyse and study each of them with care, and their relationship one with another, and all will come clear.

*Ravenstone*
*Cumnock*
*Ayrshire*

F. GRADON

## PUBLISHER'S NOTE

The illustrations, which are in the main reproduced one per page, show a wealth of information which is necessary to the maintenance engineer in the day-to-day control of his work, and its correlation with other departments of the organisation. Requirements and conditions vary considerably from one undertaking to another and, with this in mind, the illustrations have been reproduced directly from hand-prepared originals in order to demonstrate the appearance of forms prepared in this manner and how they can be adapted to suit individual needs.

*Chapter 1*

# INDUSTRIAL ORGANISATION
# and
# Elements of the Maintenance Structure

## BASIC FORMS OF INDUSTRIAL ORGANISATION

Industrial activity, in all its variety, embraces manufacturing, producing, processing, service and utility organisations or undertakings. These may be small, medium, or large single units or more complex combines and public concerns. Whatever their size and purpose, however, their organisation is built around a number of primary functions; for example, direction, management, production, sales and marketing, finance, secretariat, and legal. The extent to which these functions are segregated, augmented, and developed is dictated by the nature, size, and complexity of the undertaking concerned. As it develops, the organisation is shaped to meet the particular geographical, technical, and personnel problems and situations which obtain. Sub-division or delegation of authority becomes necessary and integrated functional branches, together with specialised and complementary services, require to be established within the general organisational framework. From this emerge the general patterns of industrial organisation referred to as line organisation, functional organisation, and staff and line organisation. These are illustrated in principle in Fig. 1.1. The actual organisation will take account of function, product, and customer orientation.

### Line organisation

This form of organisation derives from the early stages of the Industrial Revolution. It represents the first step in the evolution of industrial organisation; from the position in which the owner engaged or hired his labour and directed and controlled all operations of his business in military or autocratic style. Line organisation provided for delegation of authority with retention of control. Responsibility and authority pass directly down the lines. Unity of command is retained; each departmental head or delegated authority is held responsible for all aspects of his department's activities; each intermediate member in the line of command is involved in the full range of control functions. While the features of unity of command and clear-cut lines of authority are much in its favour, an organisation of this type can throw too wide a diversity of functions on the delegated authorities as the undertaking grows, and tends to result in the creation of 'water-tight' departments, the development of administrative sluggishness, and frustration at the lower levels of command.

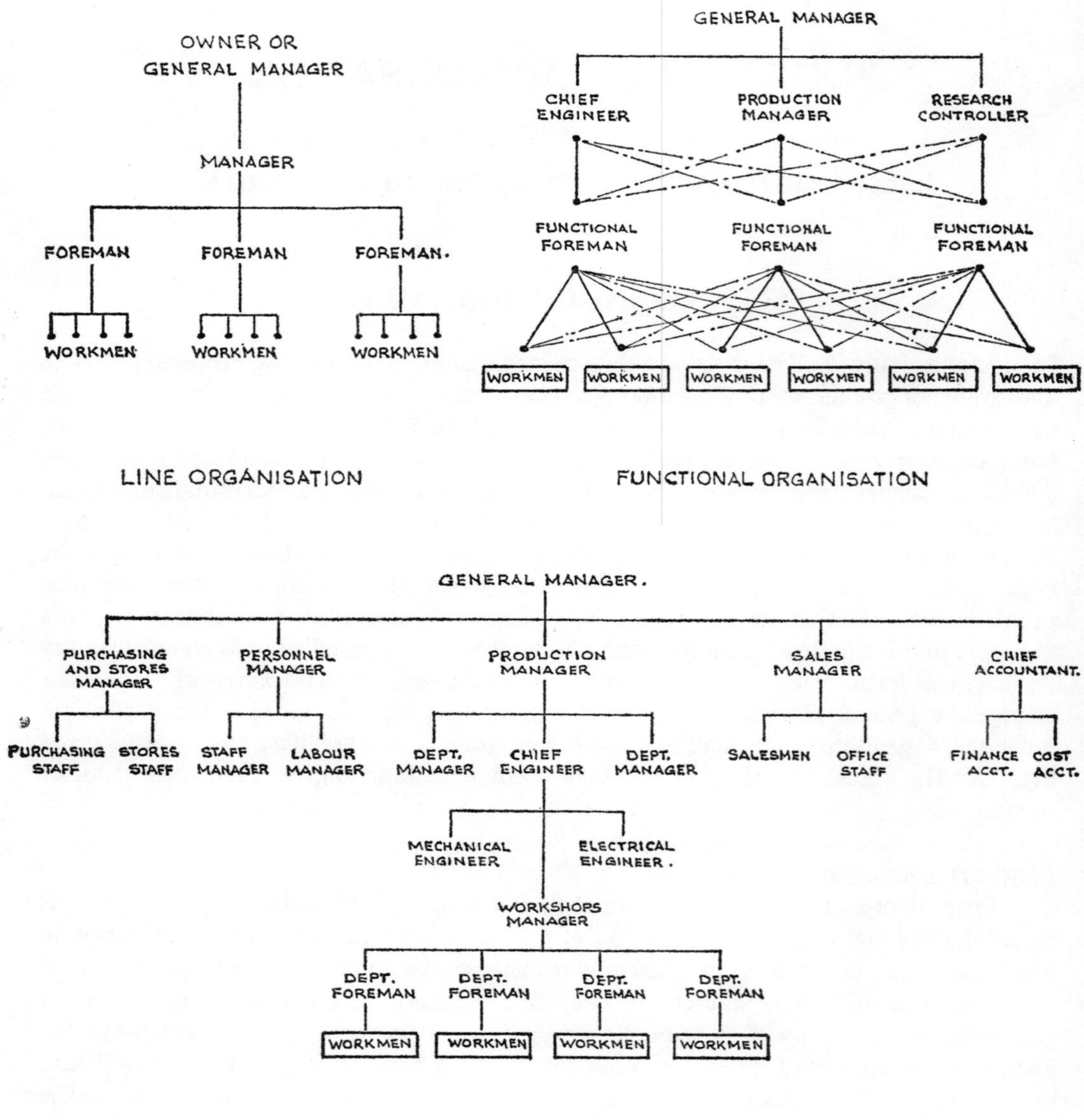


FIG. 1.1. General patterns of industrial organisation.

## Functional organisation

This form of organisation is based on the principles of establishing the various functions to be performed and grouping like functions together. Within the grouped functions, it can result in well-integrated activity, but, if developed too far, overall co-ordination becomes difficult and functional authorities are liable to clash.

## Staff and line organisation

This form accepts the need for specialists under functional heads of departments, each responsible for a particular facet of the undertaking's activities. These functional departments may be further split into functional branches, each with heads responsible to the department's functional managers. The functional departments are responsible to the line-management for advice in the formulation of policy, and for implementing agreed policies within their spheres of competence on behalf of the management. Authority flows directly down the line-management from the 'direction', through the intermediate levels to the activity level. There are off-shoots embodied in the structure at various levels where specialists exercise authority in particular fields.

It is from these basic conceptions of organisation that the forms employed in industry are developed. There is no set pattern to suit all applications and the actual structure is more often a combination of the several forms, designed to secure the optimum overall advantage from each of them, while eliminating the disadvantages.

Directive policies are of primary importance, and the responsibility for tailoring the organisational set-up to suit the nature and size of the undertaking, and for ensuring that it results in all-round operational efficiency lies with the directorate and the top-line management. Sound judgement, supported by advice arising from experience in the various functional channels, together with co-ordination and delineation of authorities, their scope and practices, are essential; otherwise top-heaviness, overlapping, confusion, inefficiency, frustration, and bureaucratic relationships are likely to occur. The calculated swing of a clock pendulum, with its compact top assembly and its correctly weighted and adjusted lower extremity, enables the clock to work with precision and efficiency. This analogy can be related to a good industrial organisation, where line-management should pivot on a small, finely balanced and co-ordinated specialist assembly, with carefully considered added weighting and widening of the functional departments and branches as they approach the level of implementation and operation. It is relevant to remember that the nearer the weighted extremity of the clock pendulum is brought to the upper pivot, the quicker the clock works, and conversely.

As growth takes place, there is a risk of over-development of the functional departments and branches at the higher levels, with each devising and attempting to impose rigid uniformity of practices and procedures within their own areas, to too great an extent. A restrictive climate is thereby created, in which there is an insistence on the observation of needless formalities and a constant introduction of new rules and procedures. This results in turn in the suppression of personal initiative, scope, and enthusiasm, upon which the organisation ultimately depends for its prosperity. A dictum attributed to Thomas Jefferson may well be borne in mind: "Error of opinion may be tolerated when reason is left free to combat it".

A good organisation gives drive, direction, purpose, and co-ordination from the top, but recognises the need to retain initiative and operational freedom at the lower levels, and so gains the advantages of the knowledge and experience which accrue from day-to-day contact with the problems and conditions at those levels. Free discussion and interchange of views and experiences enable the best practices to be sifted out. Those most suited and proved to achieve the desired results can be adopted, and those unsuited can be modified or replaced. This is a continuous process, as progress and improvement always fall short of perfection.

While conditions vary widely, there are certain essentials in an organisation which must obtain, if it is to achieve smooth and efficient working and effectiveness at all levels. Features of first importance are:

The clearest possible delineation of authorities and allocation of duties and responsibilities, taking account of geographical, functional, and personnel factors, and having regard to expediency, avoiding overlapping, and retaining flexibility.

The vertical lines of authority, or 'line-management', carrying the through thread of direct responsibility, should be as short as possible, to transmit the impact of direction most effectively. The weakening of this line by the introduction of subsidiary levels, tending to become mere transmission points for information and instructions, should be guarded against.

The establishment and maintenance of an optimum number of staff directly responsible to one individual. This must be appropriate to the level at which it applies; for example, it may provide for four to six in the case of a senior executive, but twelve or more for a supervising chargehand. The duties or type of work, the degree of uniformity, and the amount of supervision required, must be considered, to establish a satisfactory ratio in each case. A balance must be struck between waste of supervisory talent and inadequate supervision.

The consideration of individuals' personalities and ensuring that they have adequate flexibility of mind, so that within the organisational structure there is a satisfactory fusion of personnel and maximum efficiency is achieved.

Periodic reviews of performance and conditions at the operational centres of activity; ascertaining weaknesses, and considering and taking remedial actions.

## THE GROWTH PATTERN OF AN INDUSTRIAL UNDERTAKING

Having regard to later considerations it is useful to consider here the normal growth pattern of an industrial undertaking. This is outlined in Fig. 1.2. As growth takes place, off-shoots of Unit activity are separated from the central organisation. Certain administrative and functional activities remain centralised, while others become self-contained within the activities embraced by the Unit, so that the whole requires co-ordination. The organisation becomes, in effect, a group undertaking. As this process advances and depending on the nature and

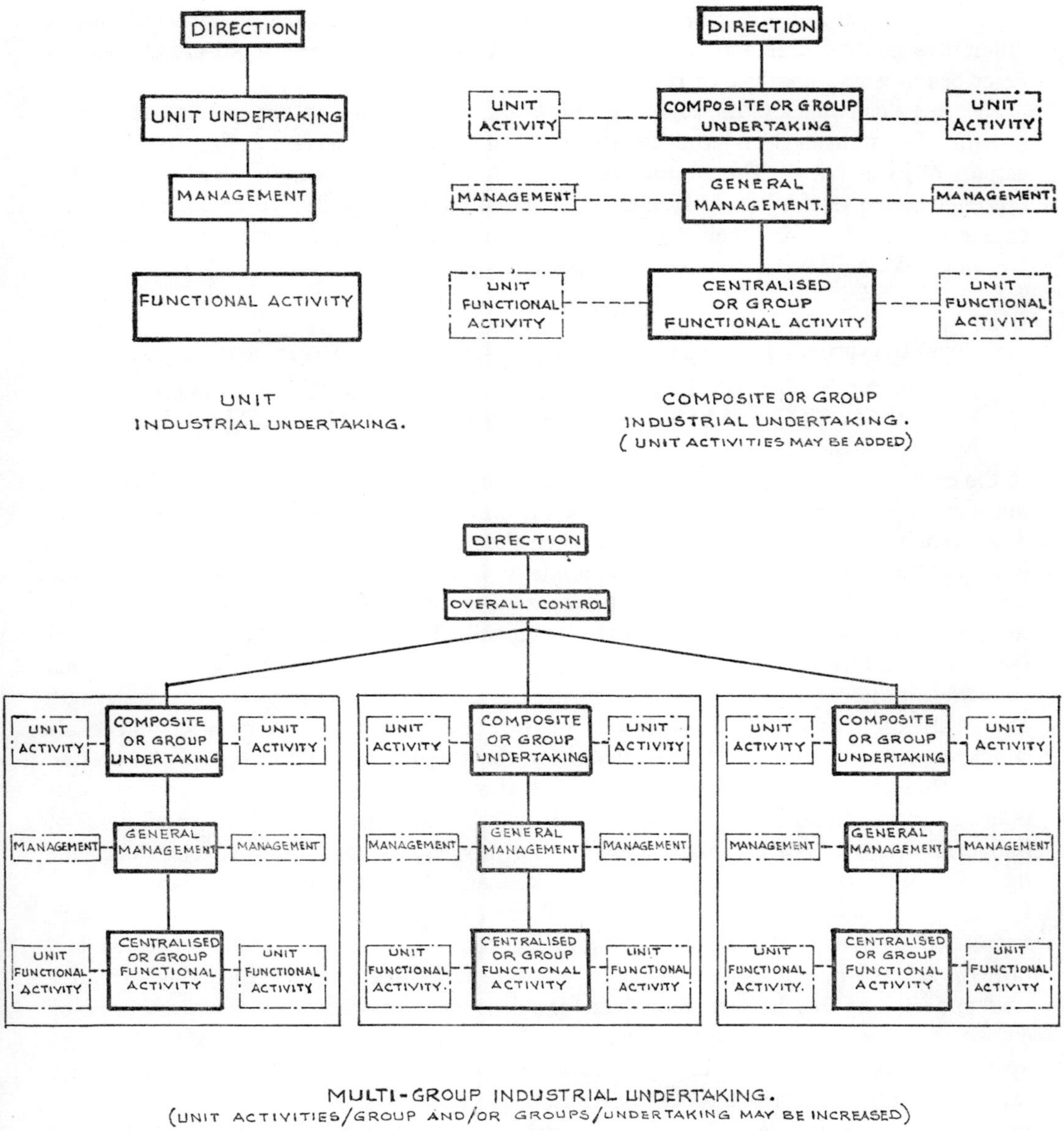


FIG. 1.2. Growth pattern of industrial undertaking.

size of the undertaking concerned, this group concept develops into a multi-group undertaking, each group being largely self-sufficient, but subject to overall centralised direction and control.

Most industrial concerns generally conform to this growth pattern, although there may be differences in structures and terminology; for example, groups and multi-groups may be described as areas, districts, or regions.

Relating this pattern to the subject of this book, whatever the nature and size or stage of growth of an industrial undertaking, there will normally be a service engineering function as part of its activities. The administrative and functional activities involved in providing this engineering service have characteristics common to all cases. The co-ordination of these engineering activities and their integration within the main organisation must be attuned to the undertaking's development, at all stages of its growth.

## THE DEVELOPMENT OF THE SERVICE ENGINEERING FUNCTION IN INDUSTRY AND ITS INTEGRATION INTO THE MAIN ORGANISATIONAL STRUCTURE

In the early stages of the Industrial Revolution, engineering service maintenance fell to the lot of enginewrights and millwrights, who learned the hard way, and had to cope with breakdown repairs of any description as and when they occurred. Labour was plentiful, wages low; plant and machinery were simple, crude and heavy; facilities were very limited, and brawn and muscle played no small part in operations. Such conditions provided great scope for the development of the initiative and versatility of those in charge and from the ranks of these men emerged many of the professional engineers of those days. As time passed, great changes came about. Sources of motive power previously confined to the muscular strength of men and animals, the elemental forces of nature and the expansive force of steam became supplemented by the introduction of electricity, the internal combustion engine, compressed air, and developments in fluid power and hydraulic transmission. These developments opened the way to a rapid and continuing growth of technological progress in all fields of industrial activity. New materials, methods and processes were evolved. Machines became more elaborate, complex and productive. Craftsmen became a feature of the labour scene. In the last decade or so, mechanisation of processes and automation of whole plants, stimulated by rising labour costs and growing competition, have rapidly advanced and brought about a revolutionary change in the industrial climate. The smaller type of industrial undertaking has become more machine-dependent and complex while the growing emphasis on larger units, group mergers, conglomerates, and public undertakings, brought about by economic pressures and governmental policies, has presented new problems—not least in the administrative and control fields.

Maintenance departments have found it increasingly difficult to cope satisfactorily with servicing under these more sophisticated conditions. Heavy expenditure on production and service plant and equipment demands a sustained high level of performance and utilisation, and this has become recognised as a vital factor in productivity and profitability.

Maintenance activities can no longer be regarded simply as servicing and breakdown repairs, carried out as required. They must be included within a comprehensive, well-organised and administered engineering service, under trained technical supervision, operating as a part of the undertaking's main organisational structure and its overall functions. Maintenance engineering has become an important element in the running costs, and hence the profitability of an enterprise. In larger undertakings the senior maintenance staff should have professional qualifications, and the chief maintenance executive should have managerial status. Maintenance work must be planned and organised with a view to avoiding breakdowns, extensive repairs, and replacements. It must be concerned not only with sustaining a high economic level of performance and utilisation of plant and equipment, but also with making its contribution in the fields of plant and machinery design, selection, application, and operation. It must help to preserve and improve safety standards and also accept accountability for projected and actual costs within its sphere of operations. Personnel, human relationships, and training all come within its range.

In the administrative area maintenance has become a major consideration of the overall management and control of any successful industrial undertaking. In the case of the larger and more complex group undertakings it is common for the engineering service organisation to be centralised, with functions tailored to suit the day-to-day operation and control of the engineering services at the separate plants or Units. The centralised service provides for close liaison with other functions of the main organisational structure of the undertaking, and embraces facilities which could not otherwise be provided at the separate Units. It can relieve the latter of much of the actual repair and overhaul work, leaving the site staffs to concentrate more on inspections made to planned schedules, detection and anticipation of weaknesses, scheduled routine short-term maintenance and servicing, and 'first-aid' repairs.

However, whether the nature and size of the undertaking calls for centralisation on a composite, group, or multi-group basis, or more localised service for a single Unit (refer to Growth Pattern—Fig. 1.2), the maintenance activities will advisedly be so organised and carried out that they provide for:

Overhauls and repairs on a planned basis, at the workshops, of a regular flow of mobile or transportable units of plant, components, and assemblies.

Periodic overhauls of fixed plant and equipment on a systematic basis by site personnel, augmented from the workshops or by external contractors when necessary.

Day-to-day scheduled maintenance on site.

Breakdowns and emergency repairs.

In addition the maintenance organisation:

Facilitates compliance with all statutory requirements, internal directives, and mandatory instructions.

Contributes effectively to safety, minimising of breakdowns, reduction and deferment of expenditure on replacements and new plant; these aims are attained by a high general economic standard of plant upkeep and reliability.

Highlights weaknesses of design, and abuse and neglect in operation; facilitates close investigation into these matters, based both on knowledge and experience.

Has the means of assessing maintenance costs and accepting accountability for them.

Provides as good an economic engineering service to production as possible.

It should also be able to call on specialist services in matters such as research into design and metallurgical failures; study of methods and practices; and the checking of the efficiency of plant performance.

Irrespective of the size or stage of growth of an industrial undertaking, there will be a service engineering function in its overall activities. This service function will normally involve design, manufacture, installation, operation, and maintenance. Whatever the nature of the plant and equipment of an undertaking, the physical processes of maintenance provided by this service function will normally include cleaning, lubrication, painting, inspection, servicing and running repairs, overhaul, improvement, and replacement.

According to the stage of growth of the undertaking the means of fulfilling this service function will normally require the following, either as unified activities, or as separate correlated activities:

Engineering office and maintenance control.
Plant record and inventory.
Drawing office.
Workshops.
Site maintenance establishment.

Fulfilment of the service engineering function is dependent on other functional departments within the undertaking, with which there must be close co-operation; these are usually purchasing and storekeeping; transport and communications; costing and finance; and general administration.

It is now possible to view the subject of maintenance in clearer perspective. It is apparent that there are functions directly involved in organised maintenance, as part of a total engineering activity, and other functions which impinge or bear upon it and must be taken into account if it is to be effective and efficient. These include:

## *Engineering functions*

Design is influenced by and requires to take account of maintenance.
Maintenance may involve supportive manufacture.
Installation and operation have a direct bearing on maintenance.
The plant record and inventory is fundamental to the control of maintenance.
The drawing office provides an operational service to maintenance, as distinct from design.
Workshop design, lay-out, organisation, operation, and control must take account of the needs of maintenance.

## *Related functions*

Production demands the services of maintenance.
General administration includes the human aspects of maintenance.
Transport and communications provide a service to maintenance.
Purchasing and storekeeping are closely associated with maintenance requirements.
Maintenance costs and accountability are important financial considerations.

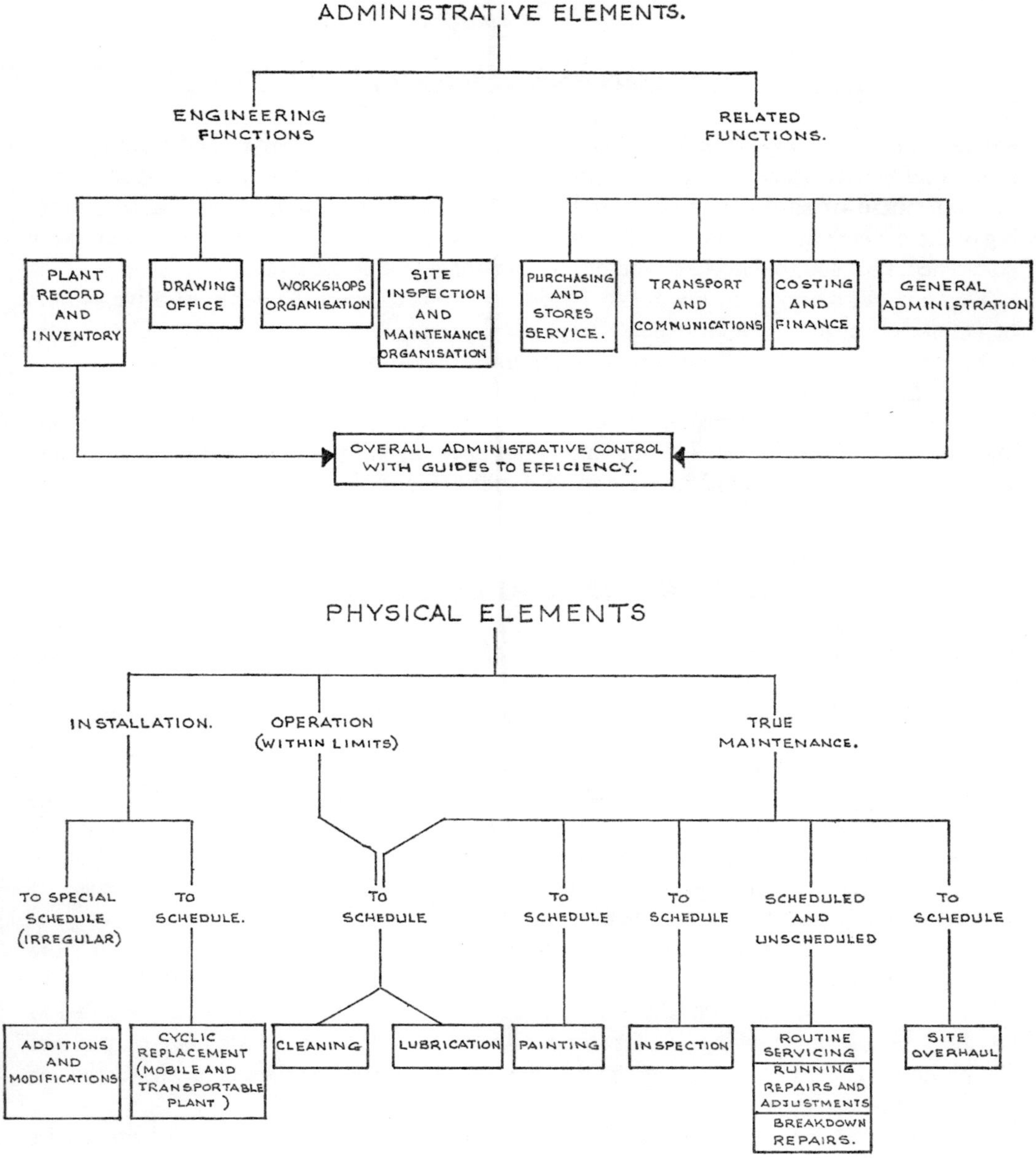


FIG. 1.3. The total maintenance service concept.

This at once emphasises the necessity and importance of a co-ordinated approach to maintenance administration, if its practical accomplishments are to match the demands made of it.

## THE TOTAL MAINTENANCE SERVICE CONCEPT

Against this background it is possible to establish the concept that organised maintenance, in its total service sense, includes two classes of basic elements, namely administrative and physical (these are shown graphically in Fig. 1.3). It is from the examination and conditioning of these elements and their integration into the structure and economy of the undertaking concerned, that sound, effective, and efficient maintenance is achieved.

This task is attempted in this book. Its purpose centres upon the administrative aspects of maintenance, upon which depend its physical processes; the latter are considered only to the extent necessary to relate them to administrative organisation and control.

As a basis from which to develop the subject it is appropriate to apply the above concept to the organisational structure of a typical industrial undertaking, highlighting the engineering services. A group industrial undertaking has been taken as offering the widest coverage and flexibility, having regard to the growth pattern (Fig. 1.4).

## ADAPTATION AND INTEGRATION

Adaptability is an essential feature of any work of this nature, and a few notes on this aspect are given here.

Taking first the administrative functional elements as explored in the following chapters:

> The principles involved in the establishment of a good plant record and inventory are essentially the same, whether it relates to 500 or 20 000 items, and whether it is operated and controlled by the works engineer himself, or is a sub-section of the engineering establishment. The system here described is comprehensive, but its scope can be narrowed; for example, it may not be politic to include financial data as a service to the finance department. A system is more readily pruned than expanded.
>
> The basic principles of good drawing office practice and operation apply, whether the works engineer performs and controls the function on his own, or whether there are 50 draughtsmen and staff. The description of a drawing office is in no sense intended to be a treatise on drawing office practice. As such, it would be inadequate and out of context. It seeks only to bring the drawing office function into focus as an element of maintenance administration.
>
> In any maintenance organisation, workshops of one form or another will have a part to play. If they are best to serve their purpose, their internal practices must not only be sound in themselves, but geared to the main organisation of which they form a part. The section under this heading aims to evaluate the factors arising in establishing, organising, and operating a service workshop, as distinct from a production department.

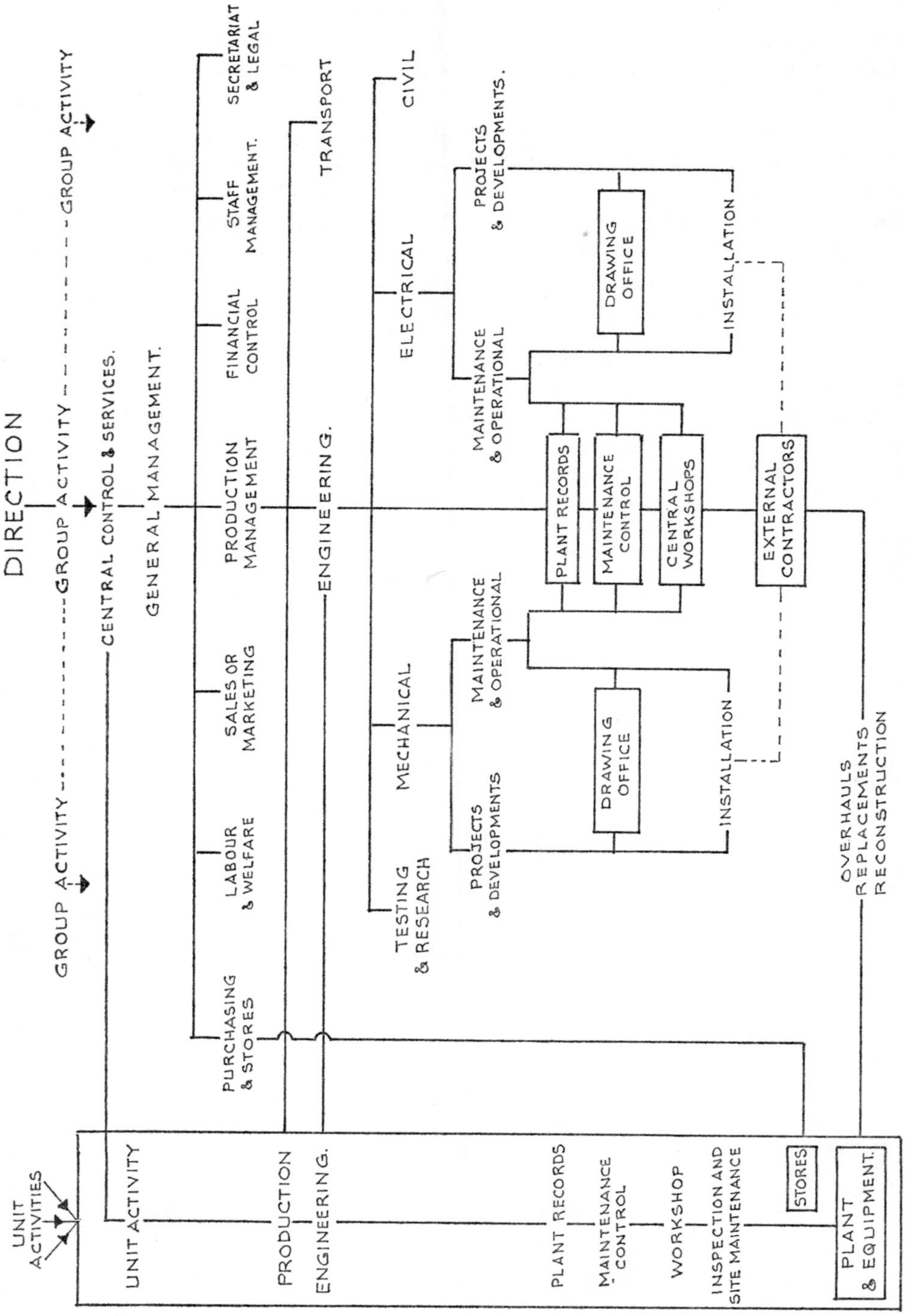


FIG. 1.4. Organisational structure—composite or group undertaking—highlighting the engineering services.

It is based on a scale of throughput and variety of work that is related to a large and complex industrial undertaking. The principles, however, can be adapted to smaller undertakings, and the scope scaled down to suit; for example, it may be that the planning and progress operations, and work-routing need not be a centralised function, but may be operated by the shop foremen and chargehands.

The site inspection and maintenance organisation described is built around the physical elements of maintenance (*see* Fig. 1.3) which can be regarded as normal. It involves work schedules and man-power deployment, which are basic ingredients of organised maintenance. Being Unit based, and developed on a cumulative 'zone' system, each of which envisages from, say, 10 to 25 maintenance personnel, this section is readily adaptable to a wide size range of organisations; similarly with the remaining functional elements.

The documentation and procedures are developed to co-ordinate the activities under the various functional elements in the organisational structure of the selected typical industrial undertaking. The variances between that structure and any other given structure will give a lead to the adoption and modification of the necessary documentation and procedures.

*Chapter 2*

# THE PLANT RECORD AND INVENTORY

## Organisation, Establishment and Control for Maintenance Engineering and Allied Services

In considering the organisation, operation and control of planned maintenance, one of the first considerations is the establishment of a sound plant record and inventory system. It is fundamental to the maintenance function, and holds the key to the solution of many problems arising in the evolution and control of a composite overall scheme. It merits careful study, allied with an understanding of the conditions in which it is to operate, and the demands which will be made upon it. If correctly devised, it performs many functions and makes a substantial contribution to the success of maintenance as a whole and thereby to the primary factors of safety, production, and economy.

It is well to start by defining the terms:

*Plant Record* has a singular basis and refers to the record and history of a single plant item.

*Plant Inventory* has a plural basis and refers to a quantitative and descriptive summary of plant items.

The scheme is basically a central service located at the Group's headquarters, but is necessarily linked with the records at the Units. Both mechanical and electrical plant are included. Having regard to the fact that in certain categories of plant (such as a pump or a compressor) the electrical portion may be separated and replaced by other equipment, it is sound policy to keep separate the mechanical and electrical sections of the record and inventory system, and to cross-reference them in such a way that the inter-relation of the mechanical and electrical components of plant items is readily established.

The system should be evolved around the following functions and requirements:

It should be designed to identify all existing and incoming machinery and plant, and to provide:

Technical data for the Group's engineering activities under the production department.

Financial data for the Group's finance department.

Purchasing information to meet the requirements of the Group's purchasing department.

Spares codings in accordance with the Group's stores procedures.

Location and movement data, to meet the needs of production, finance, purchasing and stores services.

To be effective and efficient, it must ensure that:

Account is taken of all existing plant and machinery (from the date of inclusion in the system) and of new machinery and plant (from the confirmation of the order) to the date of its eventual disposal.
Account is taken of the operations of the Group's workshops and stores, including the development of planned maintenance in liaison with the Units.
Links are established with the Group's engineering technical control and drawing office.
A clear-cut procedure, with well-defined responsibilities and duties, is established, to keep the records up-to-date; the whole backed by the authority of the Group's management, to be observed by all concerned.

## ESTABLISHMENT AND CONTROL

### General organisation

As the system is designed around the dominant and wider requirements of the Group's engineering organisation, serving production, the control should be vested in the Group's engineering department, with the provisos that the necessary liaison is established with the Units' records; that close collaboration is maintained with the finance, purchasing, and stores sections; and that the resulting establishment provides a satisfactory service to all departments and functions.

## THE CODING SYSTEM

### General category identification

From the engineering point of view, and since much of the plant is mobile and transportable (a feature to be exploited to the full in planned maintenance), it is convenient to establish plant identity basically around the type or category of plant. A code of index letters for the various general categories of plant involved should be drawn up. It is helpful if the index letters themselves give an immediate visual indication of the plant category, and this can be achieved by adopting a three-letter coding; for example:

| *General Category* | *Coding* |
|---|---|
| Air Receiver | ACR |
| Boiler | BLR |
| Cement Mixer | CMR |
| Compressor | CPR |
| Conveyor | CVR |
| Pump | PMP |

### Specific category identification

Within each general category of plant, there will be different types. For example, a conveyor may be a belt conveyor, a plate conveyor, a wire-rod conveyor, or a wedge-wire conveyor. This may be allowed for by means of a prefix figure to the general plant category; for example, 2/CVR denoting a plate conveyor.

### Unit identification

The plant unit identities can be established by figures following the index letters, in numerical sequence, in the order they are recorded; for example, 2/CVR/15. The electrical portion should be separately identified and accordingly marked under suitable codings.

It is essential that each plant item should be marked in such a way that its identifying number is not readily defaceable or removable, and in a convenient position for ready observation. Metal plates, of standard design, are sometimes employed, secured by special punch-in rivets inserted into prepared holes. A more modern method is to use plastic strips with a stamping tool. If the method employed involves preparatory drilling for attaching the plates, the location must be approved as to position, particularly in the case of flameproof electrical equipment.

## THE TECHNICAL DATA

The technical reference data relating to plant and equipment can be classified as information which comes under headings applicable in general, or which applies to the specific category of plant considered. In the first case, the headings will generally be Description, Maker's Name, Maker's Number, Year of Make, Maker's Drawing Numbers, and Group Drawing Number; these being readily printed on the standard record card. In the second case, the headings which highlight the technical data will vary. For instance, a pump record must bring out clearly the type, number of stages, gallons per minute, total head, RPM, absorbed HP, efficiency, suction branch diameter, delivery branch diameter, and details of fittings. A locomotive record must give the type, rail gauge, wheel base, working weight, tractive effort, overall dimensions, and general particulars of main components. It is therefore useful to devise a suitable lay-out of data headings relating to a particular category or categories of plant, to be entered in the 'Technical and General Data' space on the standard record card. Undue elaboration of the technical data by such design details as shaft diameters, gearing details and so on, should be avoided. The plant record office should not attempt to take over the function of the drawing office, and cross-reference to the drawing office records for design details should be the general practice from the point of view of both convenience and reliability.

## INTERCHANGEABILITY

This feature can be indicated by recording, in addition to the plant number of the particular item to which the card relates, the plant numbers of plant common to it; with such a reference the location or any other details of the latter can then be obtained.

## RELATED PLANT

Reference is made earlier in this chapter to the advisability of keeping the mechanical and electrical sections of the record system separate and cross-referenced. This can be achieved by recording the plant numbers of the related electrical equipment on the mechanical card, and those of the related mechanical equipment on the electrical card. This related plant record can also be used in such cases as a boiler record card, with a related plant number giving access to the records of the feed pump and its particulars; or a compressor card, with related plant numbers locating the records of the aftercooler and the air receiver, together with their particulars.

## MAINTENANCE CONTROL

Direct assistance to maintenance control can be afforded by a coding which gives rapid access to the maker's catalogues, maintenance instructions, parts lists, and lists of recommended spares; and by listing the overhaul period, and noting the forward overhaul scheduling.

## SPARES AND RENEWALS

A simple cross-reference on each record card, relating the plant number to the appropriate stores coding for its spares and renewals records, will meet the requirements under this heading. This can be shown as the stores coding reference.

## PURCHASING AND FINANCIAL DATA

The inclusion of certain purchasing and financial data in the records can be a useful service to the appropriate departments when they wish to quote previous order numbers, trace specifications, date plant, and carry out stocktaking and auditing. Suitable headings for recording the information are Purchase Date, Order Number, Date Capitalised, Rate of Depreciation, Current Value, Price and Writing-off Date. By signalling the placing of an order and the due delivery date on the record card a useful aid is given to progress chasing.

## LOCATION AND MOVEMENT

The incorporation in the plant record and inventory system of a satisfactory method of recording and controlling plant location and movement requires careful study, with a clear understanding of the requirements to be met and the problems involved. First, the existence of the following basic plant categories must be recognised:

*Fixed plant*

Stationary plant installed permanently at specific locations; for example, boilers, economisers, fans, turbines.

*Mobile or transportable plant*

Plant units of the stationary type, periodically replaced and removed for overhaul; for example, certain types of compressors, pumps and fans.

Mobile plant units; for example, locomotives, wagons, lorries, cars, bulldozers, excavators and loaders.

Single plant units which may be separated from the fixed or mobile units; for instance, motors, starters, pumps, oil coolers and fans.

For each such unit—given that the records exist from the date of ordering—it is next useful to consider its location and movement. It may be:

On order, with delivery pending.
Held at a suitable location following delivery.
Installed, or otherwise put to use at a location within the organisation.
Withdrawn from use for repair or overhaul on site, or at the organisation's central workshops, or at an external firm's works.
Replaced into service following repair or overhaul.
Held as reconditioned plant.
Withdrawn from service and classified as redundant, obsolete, or scrap.

## OVERHAULS AND RECONDITIONING

Two further aspects should be incorporated in the plant record system; namely, forward forecasting of plant overhaul requirements, and a record of all overhauls executed.

The plant record system is primarily a Group records function, as much of the overhaul and repair work is related to the capacity of the central workshops, which are also a Group function. The main risk is of unnecessary complexity and the policy should be to provide a root source of information, from which detailed information can be gathered by personnel and departments according to their particular needs.

For instance, the forward forecasting of overhauls is sufficiently catered for by a broad programme of monthly commitments on, say, a three-year basis; this provides a guide to Unit and central workshops overhaul loads, and their execution on a pre-planned basis. This is shown in the records by inserting the overhaul period and frequency, and signalling the relevant dates. This provides a practical basis on which the replacements and overhaul requirements of Units can be co-ordinated with the Unit and the central workshops, and also for an efficient organisation.

With regard to the record of completed overhauls, it is sufficient to cross-reference the overhaul job numbers. This provides immediate reference to the appropriate completed job record card in the central workshop's planning department.

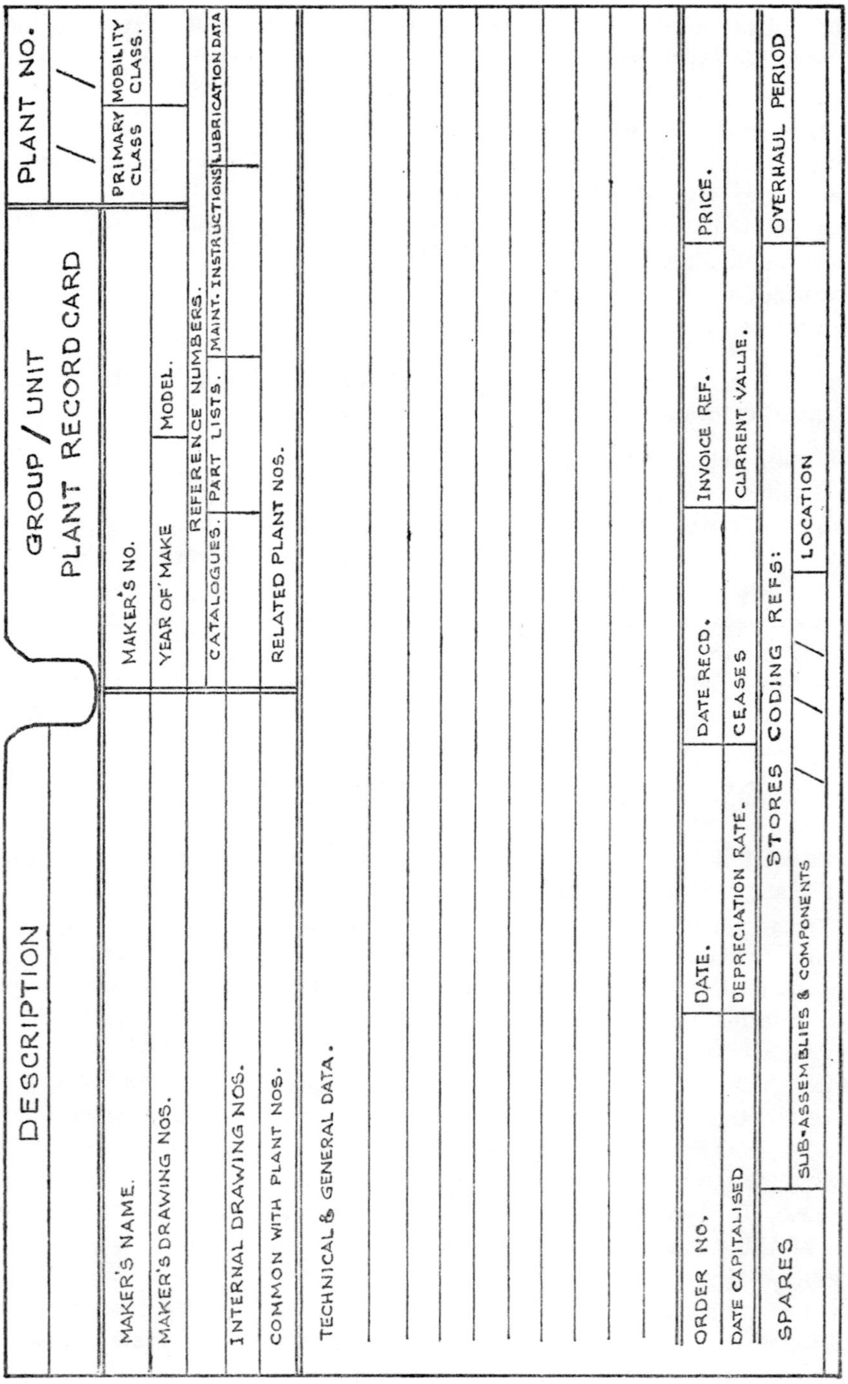


FIG. 2.1. Group/Unit plant record card.

## THE RECORD CARDS

To illustrate how the Group's system requirements are met, the data and information relating to a particular plant item are shown in Fig. 2.1 (the Group/Unit plant record card) and Fig. 2.2 (the Group/Unit plant history card). The latter are inserted in the record cabinets such that the lower edges form a visible index; selected coded colour signals are used to highlight the required information.

The relative Group/Unit plant record card is placed in the rear of a folder which also contains the plant history card.

Thus, all the recorded data for each plant item are presented in readily assimilated visual form.

The records of a particular Unit are duplicated at that Unit. This eases liaison between the Unit and Group maintenance operations, and assists the Unit's operating staff with day-to-day work; for example:

It gives immediate cross-reference to the Group's plant records system.
It provides basic technical data on the Unit's plant and equipment.
It gives full access to the Unit and Group stores coding system for spares and renewal stocks.
It enables the operating staff clearly to identify and specify plant requirements, with the minimum of manpower, time and risk of error.
It forms a useful adjunct to, and extension of, the planned maintenance control records.

When plant is withdrawn from the Unit (for example, mobile plant for periodic overhaul, replacements having first been provided), the Unit's record cards accompany the plant movement and receipt advice (Fig. 2.4) to Group's records. The Unit's record cards are subsequently returned by Group Records, with a plant movement and receipt advice, when the overhauled plant is returned to the Unit. In the case of transfer of plant directly from one Unit to another, the dispatching Unit sends the record cards together with a movement advice to the receiving Unit, and the latter sends a movement advice to the Group's records office. This serves to keep the Unit and Group records up-to-date.

## THE RECORD CABINETS

A wide range of standard visible index cabinets is available; when choosing cabinets and deciding on card sizes, the makers and stockists should be consulted. The cabinets have standard heights and depths. A typical 16-slide cabinet, with 63 cards per slide, will house the cards for 1000 plant units.

The following points should be borne in mind when choosing card index equipment:

It should be of the visible-index type.
Metallic swivel roller mounting of each card, and easy removal and replacement of any card, are desirable features, bearing in mind the permanent nature of the equipment, the frequent 'flipping-over' of the cards for reference, and the substitution of cards to keep the records up-to-date.

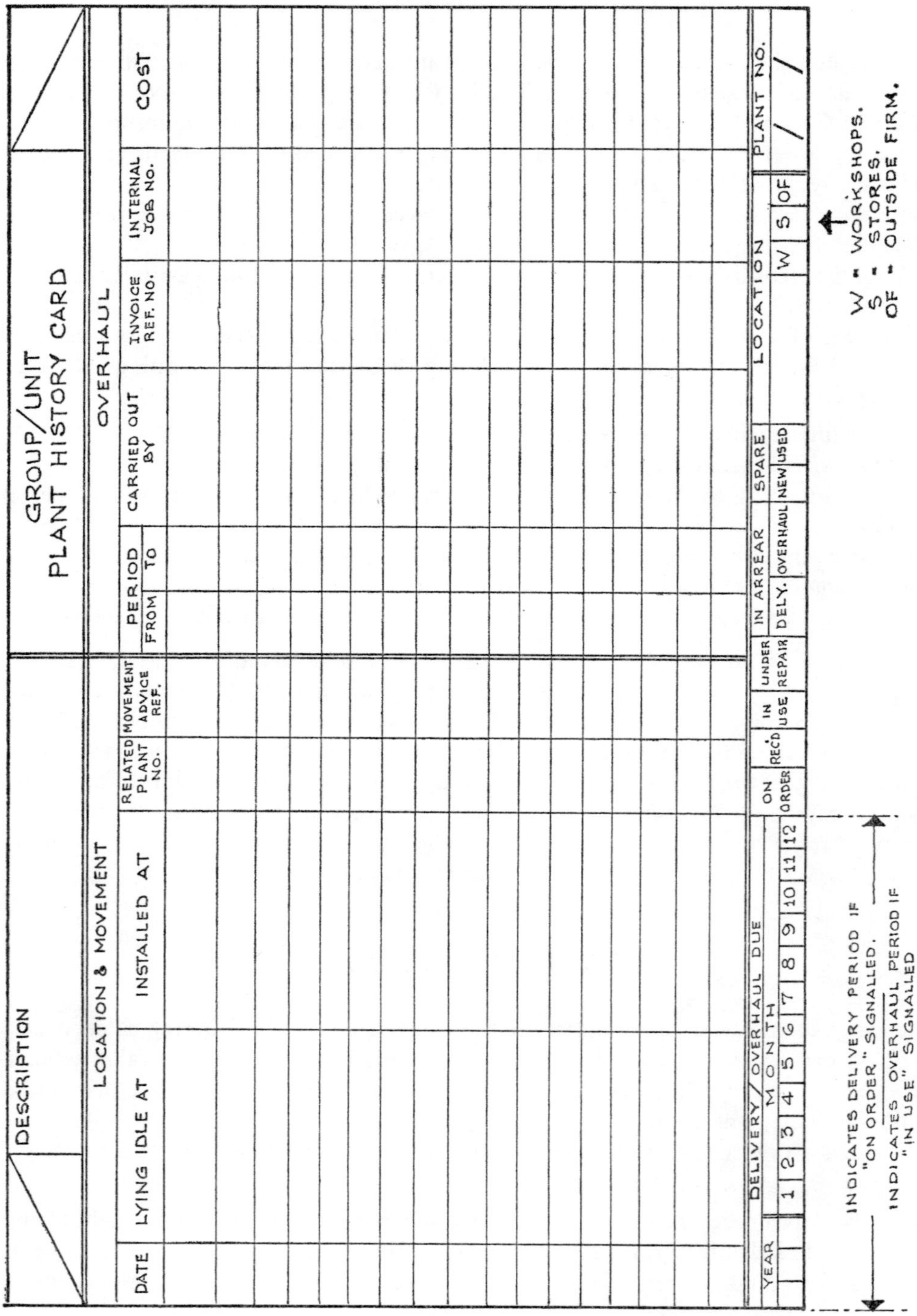


FIG. 2.2. Group/Unit plant history card.

Each slide should withdraw to a fixed stop, and then be free to hinge downwards to give convenient reference; the signals on the cards should be easily seen and adjusted.

The index signals, which are available in various colours and designs, should be chosen for their clarity and durability. They should slide into plastic strips fixed to the bottom of the flaps; these preserve the cards against continuous fingering and guard against inadvertent displacing of the signals.

Provision should be made for the insertion of bold labels on the front of each slide, so that segregation of plant categories is clearly displayed.

The cabinets should be supported on tables at a height giving convenient access to all the slides contained.

## OFFICE ACCOMMODATION AND LAY-OUT

The Group records office is the hub upon which the many functions of the plant record and inventory system revolve. A logical approach to its establishment and accommodation may be made by:

Establishing the existing and probable peak numbers of plant items and equipment to be recorded.

Establishing, from the proprietary systems available, the standard size of cards to be adopted.

Selecting the cabinets to be used, with provisions for segregating the plant and equipment into categories, and for signalling data on each card.

Establishing the number of cabinets required, and their lay-out.

Determining the other office furnishings and equipment required, and the lay-out most suited to the routine and functions to be fulfilled.

The number of plant items and equipment to be recorded is derived from the inventory build-up in the various categories, described under the next sub-heading. There are wide differences between small industrial undertakings, group undertakings, and multi-group undertakings; these, together with the variations in the nature of their plant and equipment, have a large effect on the scale and range of the plant record and inventory, which may range from a few hundred to many thousands of items. Provision should be made for a future extension of, say, 15 to 20% of current capacity.

An enlarged signal key to the index in the records should be prominently displayed, to help to interpret the system.

The general furnishings and equipment will include desks, tables, chairs, typewriters, and provision for housing the inventory summary files, and for stationery and records incidental to the operation of the system.

Figure 2.3 illustrates an arrangement of a typical records office. It should have good lighting, heating, and ventilation. Fireproofing and fire precautions are major considerations.

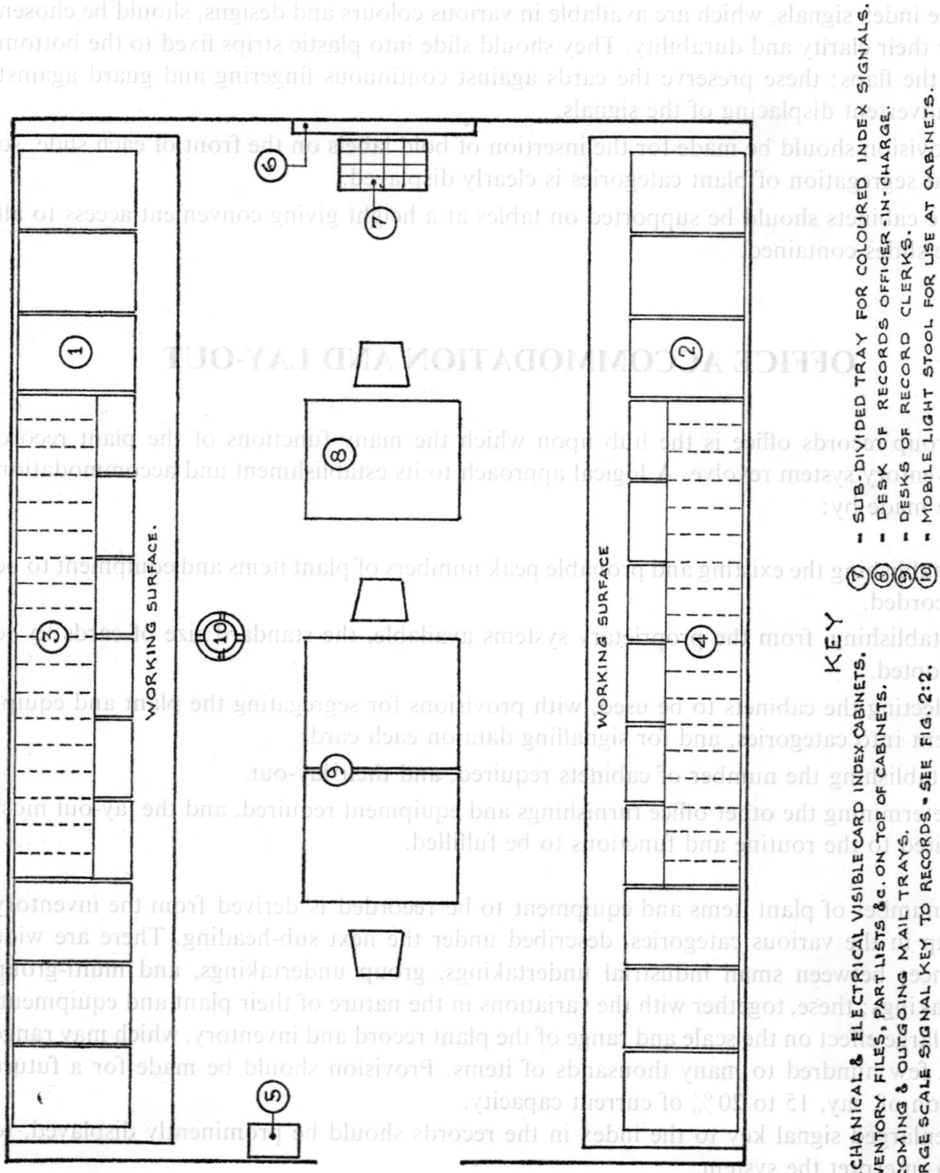


FIG. 2.3. Plant record and inventory office lay-out.

## COLLECTING THE INITIAL DATA

In making up the plant inventory, together with the collection of the data for building up the plant records it is advisable to ensure that the personnel engaged on this work have a good engineering background, and are familiar with the types and functions of the plant and equipment being assessed; and that the data are so presented that they are intelligible, accurate, complete, and capable of easy assimilation into the plant record system.

The personnel engaged in the operation will depend on the extent of the work, whether it is on a Unit or Group basis, and the time available for collecting the data, having regard to the desirability of its steady and continuous assimilation into the Group's records by the staff employed there.

It is desirable to arrange that the data are collected by one or more teams, each of two persons, one of whom obtains the particulars from the plant, and the other enters it on sheets provided for the purpose.

## RELATED DOCUMENTATION

The following illustrate the further documentation necessary to fulfil and control the functions provided for in the system and are self-explanatory:

Fig. 2.4—Plant Movement and Receipt Advice.
Fig. 2.5—Forward Advice of Deliveries Due.
Fig. 2.6—Forward Overhaul/Replacement Programme.
Fig. 2.7—Weekly Return of Plant Movement and Receipt.

## OPERATION AND CONTROL

From this point the means exist of fulfilling the many functions of the plant record and inventory system described at the beginning of this chapter. It remains to consider the day-to-day operation and control; this is readily done by preparing organisation and procedure charts, showing the whole of the administrative framework in which the system operates, and the flow of documentation already described. The procedures relating to the plant record and inventory system are difficult to portray in a single chart; it is therefore desirable to sub-divide it into a series of charts, each showing one or more aspects of the system. For example:

*Chart I* Existing Plant: Collecting data, Inventory and Record System, Identifying Plant.
*Chart II* New Plant on Order: Progressing.
*Chart III* New Plant: Receipt and Inspection.
*Chart IV* Plant Receipt, Movement, Location, and Purpose of Transfer. Forward Overhaul Programme, Overhaul Records.
*Chart V* Plant Disposal (external).
*Chart VI* Financial Control.

| CODE NO. PLANT MOVEMENT & RECEIPT ADVICE | PMRA/A-001 |
|---|---|
| TO BE SUBMITTED BY BOTH DESPATCH & RECEIPT POINTS. RETAIN CARBON COPY. | |

| | FROM | | | TO | FOR RECORDS OFFICE USE | | | | | | |
|---|---|---|---|---|---|---|---|---|---|---|---|
| | (DELETE AS NECESSARY) | ISSUED BY | DATE | DELETE AS REQD. | REC'D BY | DATE | RELATED PMRA. NO. (FOR RECEIPT) RECEIVED BY | DATE | RECORDS ENTERED BY | DATE | ENTERED IN WEEKLY SUMMARY NO |
| GROUP | CENTRAL STORES.<br>CENTRAL WORKSHOPS<br>MAINTENANCE CONTROL | | | GROUP PLANT RECORDS | | | | | | | |
| UNIT<br>......<br>...... | PLANT RECORDS.<br>STORES.<br>MAINTENANCE CONTROL | | | UNIT PLANT RECORDS | | | | | | | |

| MACHINE OR PLANT INCLUDED | MAKER | MAKERS' NO. | INTERNAL PLANT NUMBERS. MECHANICAL | ELECTRICAL |
|---|---|---|---|---|
| | | | | |
| | | | | |
| | | | | |
| | | | | |
| | | | | |

| | RECEIVED (DELETE AS NECESSARY) | PUT INTO USE | WITHDRAWN FROM USE (ENTER ✓ AS APPROPRIATE) | CONSIGNED (ENTER ✓ AS APPROPRIATE) |
|---|---|---|---|---|
| USED PLANT. | DATE ................<br>FROM :<br>CENTRAL STORES<br>CENTRAL WORKSHOPS.<br>UNIT STORES............ (ACTIVITY)<br>UNIT W'SHOPS......... (ACTIVITY)<br>OTHER SOURCES | DATE ....................<br>UNIT....................<br>LOCATION ................<br>.......................... | DATE ...................<br>LYING IDLE AT :<br>.....................<br>FOR OVERHAUL & RETURN. ☐<br>FOR CYCLIC OVERHAUL --- ☐<br>FOR TRANSFER ------- ☐<br>AS REDUNDANT ------- ☐<br>AS SCRAP --------- ☐ | DATE .....................<br>TA.NO.......................<br>FOR OVERHAUL TO:<br>..................(FIRM)<br>CENTRAL W'SHOPS ... ☐<br>FOR STOCK TO:<br>CENTRAL STORES ... ☐<br>TO UNIT..................<br>TO RELEASE PLANT NO ../../....FOR RETURN TO CENTRAL WORKSHOPS BY...... |
| NEW PLANT | DATE:.................<br>FROM:.................<br>ORDER NO:..............<br>IDENTITY PLATES ISSUED DATE .............. | FIRST TIME (NEW)<br>DATE .................<br>UNIT.................<br>LOCATION............. | | |

FIG. 2.4. Plant movement and receipt advice.

| CODE NO. FORWARD ADVICE OF DELIVERIES DUE | FAD/A-001 |
|---|---|
| PLANT RECORDS TO ISSUE BY 15TH. OF CURRENT MONTH FOR ITEMS DUE FOLLOWING MONTH. RETAIN COPY | |

| DUE MONTH | YEAR | FROM | ISSUED BY DATE | TO | ISSUED BY | DATE |
|---|---|---|---|---|---|---|
| | | GROUP PLANT RECORDS | | PURCHASING & STORES (ORDER SECTION). | | |

| DESCRIPTION | SUPPLIER | STORES CODINGS. MAIN CLASSN. | SUB-CLASSN. | MAKER'S CODE. | ORDER NO. | DATE | WEEK DUE 1 | 2 | 3 | 4 | 5 |
|---|---|---|---|---|---|---|---|---|---|---|---|
| | | | | | | | | | | | |
| | | | | | | | | | | | |
| | | | | | | | | | | | |

FIG. 2.5. Forward advice of deliveries due.

CODE NO.

FORWARD OVERHAUL/REPLACEMENT PROGRAMME

FOR MONTH OF.....................19.......

| FROM | | | | TO | | |
|---|---|---|---|---|---|---|
| | | ISSUED BY | DATE. | | RECD. BY | DATE. |
| GROUP | PLANT RECORDS | | | CENTRAL WORKSHOPS.<br>MAINTENANCE CONTROL. | | |
| UNIT | PLANT RECORDS. | | | MAINTENANCE CONTROL. | | |

PLANT DUE FOR OVERHAUL OR REPLACEMENT IN THE MONTH SHOWN ABOVE :-

| FIXED PLANT | | | | | | | | | | | MOBILE AND | | | | |
|---|---|---|---|---|---|---|---|---|---|---|---|---|---|---|---|
| LOCATION | ACTY. CODE | PLANT | PLANT NOS. | | WEEKS | | | | | C'PLETED (✓) | LOCATION | ACTY. CODE. | PLANT | PLANT NOS. | |
| | | | MECH. | ELECT. | 1 | 2 | 3 | 4 | 5 | | | | | MECH | ELECT |
| | | | | | | | | | | | | | | | |
| | | | | | | | | | | | | | | | |

| | FOR ·/A-001 | |
|---|---|---|
| RETURN TO (WHEN ACTION COLS. COMPLETED). | | |
| | CARDS ENTERED | SIGNALS ADJUSTED. |
| PLANT RECORDS. | ............(INITIAL) | ............(INITIAL) |
| PLANT RECORDS. | ............(INITIAL) | ............(INITIAL) |

TRANSPORTABLE PLANT

| DUE IN W'SHOPS (DATE) | P.M.R.A. ISSUED (DATE) | PLANT RETURNED (DATE). | REPLACEMENT PLANT NOS. | | DUE AT SITE (DATE) | P.M.R.A. ISSUED (DATE). | PLANT CONSIGNED (DATE) |
|---|---|---|---|---|---|---|---|
| | | | MECH. | ELECT. | | | |
| | | | | | | | |
| | | | | | | | |

NOTES:

1. ISSUE FROM PLANT RECORDS 1ST. WEEK OF PRECEDING MONTH.
2. AT GROUP, ISSUE 2 COPIES TO MAINT. CONTROL/WORKSHOPS. RETAIN COPY.
3. AT UNIT, ISSUE 1 COPY TO MAINTENANCE CONTROL. RETAIN COPY.
4. GROUP ENTRIES UNDER (i) LOCATION. (ii) PLANT CATEGORY.
5. PLANT RELEASED TO BE DELIVERED TO CENTRAL WORKSHOPS NOT LATER THAN DUE DATE SHOWN.
6. COVERING WORKS ORDER TO BE SENT TO GROUP ENGINEERING OFFICE.

**Fig. 2.6.** Forward overhaul/replacement programme.

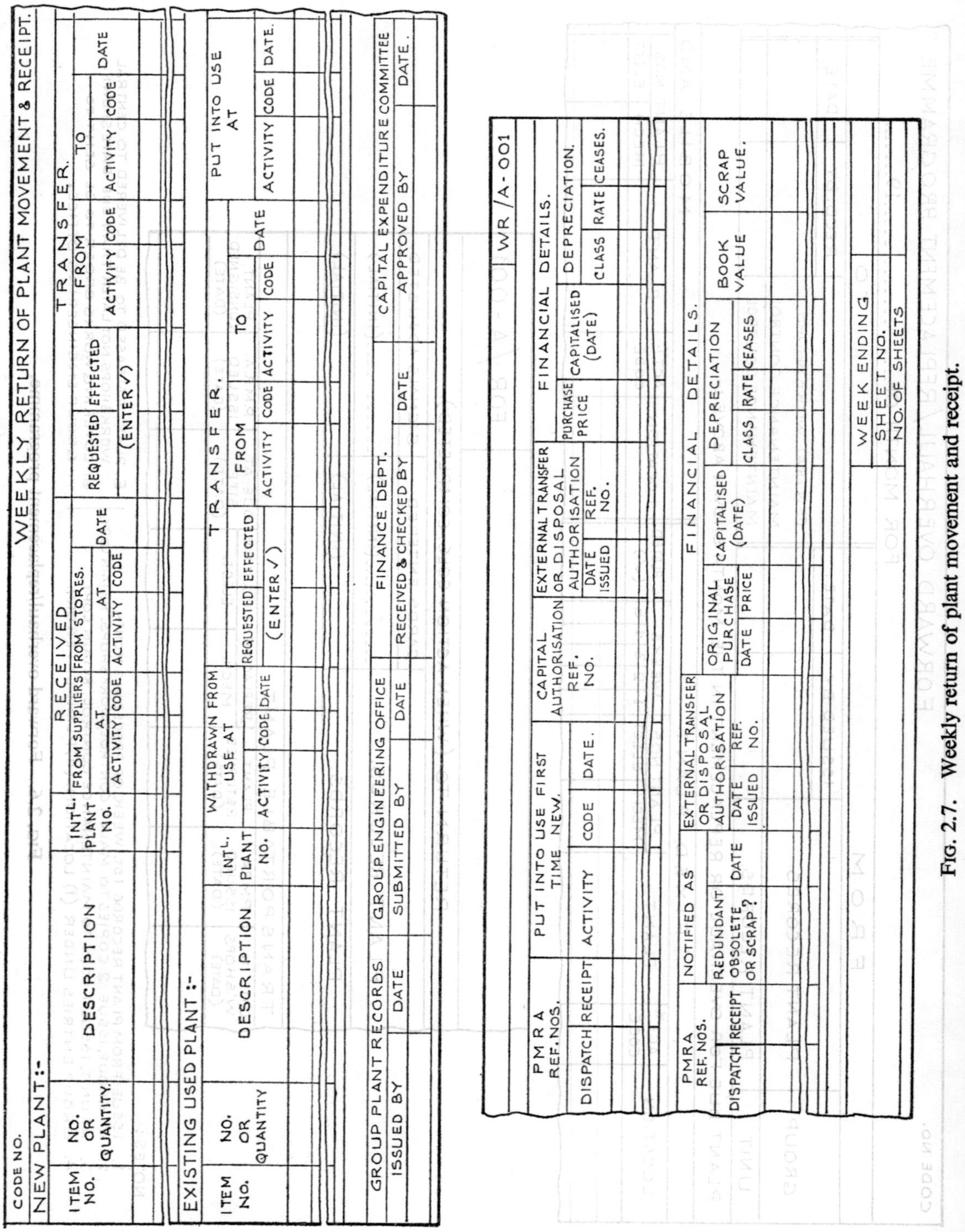


FIG. 2.7. Weekly return of plant movement and receipt.

From these, instructions for day-to-day operation at all action points can be prepared and issued. A summary of the documentation and procedure charts referred to, together with a list of code-numbered documentation, is advisedly drawn up. These should be adapted to suit the size and organisational structure of the undertaking concerned.

Lastly, two important points are worthy of emphasis. The first of these is that the records should be maintained by a neat, methodical worker, with an orderly mind and a good practical engineering background. He must insist on neatness and accuracy on the part of his staff. The second is that adherence to the established procedure by all concerned must be insisted upon.

*Chapter 3*

# THE DRAWING OFFICE
## as an Administrative Element of the Maintenance Organisation

An internal drawing office, properly staffed and equipped, is fundamental to the engineering organisation of any well-ordered industrial undertaking. It may be concerned primarily with project and development work; but since it is usually best to establish a single drawing office at the Group, it is worth while to consider its internal practices and how they can be modified to provide a useful service to the maintenance engineer. The possible benefits which the maintenance function can obtain from the drawing office are not always realised.

The drawing office is an important link between the engineering design and control, the central workshops, and the production engineering. The primary need is to get things right on the drawing board; that is where knowledge and experience are blended in a diagrammatic form for the purpose of practical fulfilment.

The drawing office is concerned with basic engineering standards as applied to design, as distinct from the standardisation of the article. The former is the logical prelude to standardisation of plant and equipment, which is a major consideration in planned maintenance and stock control. This distinction is demonstrated graphically in Fig. 3.1. Too often is the engineer confronted with the problem of standardising plant and equipment, when the basic standards appropriate to the design of the items have been ignored.

Design and specification should be closely related to the needs and conditions to be met. This bears directly on economy, performance and reliability. Close liaison must be maintained with the manufacturers of purchased plant and equipment, and with internal staffs in respect of operating experience, and the prevention of breakdowns and excessive maintenance due to faulty or unsuitable design. The drawing office influences the standard of engineering practice, which must be kept abreast of developments in materials, methods, processes, tools, and equipment. It should be alert to developments in the utilisation of high grade steels, alloys, and non-metallic materials; the application of higher economic standards of limits, fits, and tolerances; the adoption of British Standard Specifications and standards established within the industry itself; compliance with current regulations; and questions of lay-out, accessibility, lubrication, lighting, operational convenience, ease of dismantling and re-assembly, handling facilities, safety, and costs.

These considerations emphasise the effect of the drawing office on the maintenance function, and establish its importance in an overall scheme.

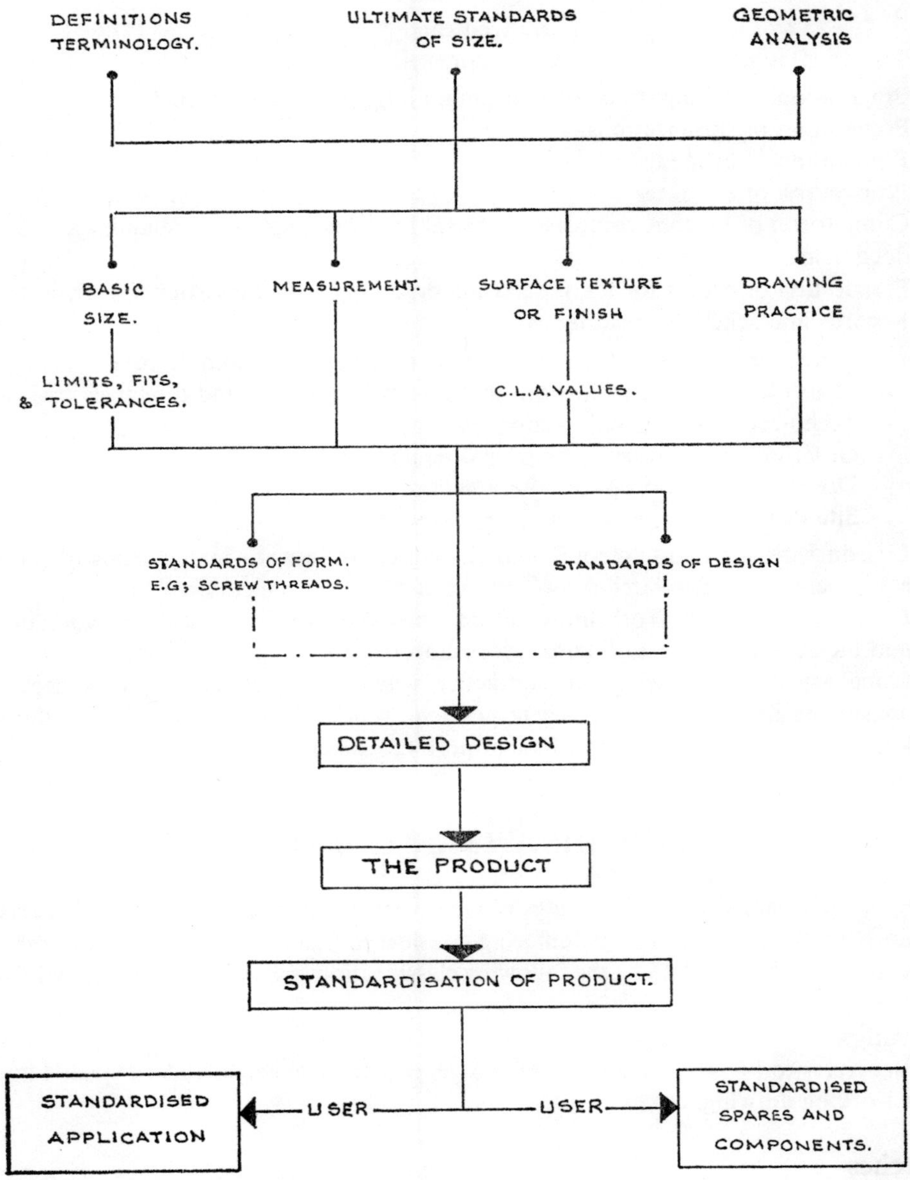


FIG. 3.1. Basic engineering standards in design leading to standardisation of product.

The functions of a drawing office may be summarised as follows:

Design: Analysis of requirements and calculations.
Lay-outs and general arrangements.
Design and detailing of components.

Preparation, checking, tracing, recording, filing, printing, and mailing of drawings.
Preparation of specifications.
Preparation of estimates.
Preparation of enquiries.
Comparison of tenders, technical appraisal and checking against enquiry and estimate documents.
Preparation of orders for the purchasing department or the Group's workshops.
Records and schedules relating to:

Drawings and sketches in existence, and new ones as completed.
British Standard Specifications and internal standards and directives.
Technical literature and catalogues.
Ordering and delivery of plant and equipment.
Drawing and sketch issues and amendments.
Site installation phasing and progress schedules.

Liaison with the Plant Records and Inventory Section, to match items of plant and equipment with the related drawings, sketches, and specifications.
Liaison with Group Workshops on drawings and sketches issued for work in hand, and incidental services to internal workshop requirements.
Some aspects of drawing office practice which relate to maintenance aspects are considered in the following. Where possible, standard practice should be adopted or adapted, in preference to introducing more radical changes.

## SIZES OF DRAWINGS AND SKETCHES

The range of accepted sizes of drawings is wide. Since it will be necessary to use at least two (possibly three) sizes of drawings, differing considerably in size, and since these should be stored and identified separately, the following classifications may be adopted with advantage:

### Drawings

Full size (refer to BS 3429), for lay-outs and general arrangements, and major and multiple detail working drawings.

### Sketches

Medium (refer to BS 3429), for medium detail and working drawings, and Small (refer to BS 3429), for minor detail and working drawings.

The larger sketch folds to the single sketch size. Drawings and sketches should be housed in separate cabinets, suited to their size, and located by means of the numbering and recording system employed.

## LAY-OUT OF DRAWINGS

Lay-out is important since it helps to ensure the inclusion of essential data and its presentation in a standard form. It can be a source of considerable exasperation to technical heads dealing with large numbers of drawings and papers, and to shop personnel, to be presented with drawings that have to be searched for information that should be immediately apparent. The lay-out of a drawing is governed by the category of engineering practice with which it is concerned, and whether it is a lay-out, general arrangement, or detail drawing.

Certain parts of the lay-out are common to all drawings, notably title blocks with name of undertaking, location, title proper, scales, date, drawing number, signatories, modifications, dates, and authority.

There are also differences; for example, a mechanical lay-out or general arrangement drawing should embody references to the related detail drawings and sketches, and (in certain cases) to composite parts lists. The detail drawings and sketches should include the relevant part lists, materials, machining and other symbols. A structural steelwork drawing should give the symbol and section relating to each component member. The structural detail drawings of the component members and fixtures should relate the latter to the appropriate symbols on the arrangement drawing.

## TITLE BLOCK

The title block should be located in the bottom right-hand corner of the drawing (when flat filing is adopted) and should be repeated at the top left-hand corner of the drawing (when vertical filing is used).

## MODIFICATIONS

When a drawing is altered, the amendments should be clearly recorded on the drawing, with dates and signatures. Lettered block panels should be provided for this purpose. An entry in panel 'C' indicates the third revision, and the drawing number should subsequently be augmented to read '. . . /Rev. C'.

## MACHINED SURFACES AND FINISH REQUIRED

The system of denoting machined surfaces and the class of finish varies widely, and it is wise to adopt and enforce a standard.

One which has much to recommend it is as follows:

| ∇ | ∇∇ | ∇∇∇ |
|---|---|---|
| Rough Machine | Smooth Machine | Ground Finish |

This may be modified to include other qualities of surface finish or process, by using a single triangle with a code letter inside it indicating the desired operation or finish, such as:

Ground Finish $\nabla$ G

The machining symbol key should be located at the bottom of the drawing for reference.
An alternative method is set out in BSS 1134 (Revised).

## MATERIAL LIST

The material list should include the number of the part or component, its description, the rough material weight, finished weight, and number required. It should consist of a ruled column block at the top right-hand corner of the drawing. In the case of stored spares, an additional column to identify the component, together with the spares coding system, can usefully be added.

## JOB NUMBER

When a job number is included, it should be shown in a ruled block above the title block.

## PLANT IDENTITY

When a drawing refers to specific plant items and equipment recorded in the plant record and inventory system, the plant number should be entered above the job number in a ruled block, to allow cross-reference between the drawing office and the plant record and inventory section.

## LAY-OUTS APPROPRIATE TO THE MAINTENANCE FUNCTION

For maintenance purposes, the following form a suitable range of sizes and classifications.

### Full size drawings

*General lay-out of works or plant (composite or diagrammatic arrangement)*
This should be a miniature reproduction in plan of the relative locations of the various components and services making up the whole. It can be used to define the boundaries of the maintenance zones referred to in Chapter 5 and the plant and services therein; to identify the items of plant within their setting; to establish correlation with the plant records and inventory; and to provide the data for related plant, in the event of replacement of a section or plant unit which depends on other equipment already in place.

### *General arrangement of a plant unit or item*

This relates to a particular unit or item, giving its general form and shape, and general information about it. It can usefully include, as appropriate:

Overall and principal dimensions.
Holding-down bolt positioning.
Weight of whole assembly and main components as dismantled, and points where weight is applied to foundations.
Reference to component assembly and detail drawings.
Plant numbers of plant to which it relates.
Maker's references and part numbers.
Stores classification references for spares and renewals.
Order references.
Related general lay-out drawing of which it forms a part, and related detail drawings.

### *Multiple component drawing*

This constitutes a further breakdown of the plant section or item into its component parts for identification and/or manufacturing purposes. It may take the form of a collection of part drawings identifying a number of components, or an 'exploded' view of a composite assembly defining the various parts and the order of their assembly, the part names and identity numbers being shown in each case. In the former case, the appropriate part drawing can be cut out, or that portion printed only, and attached to an order or enquiry for that part.

### *Major component drawing*

This serves to show in detail a single major component of a plant section or item. It should include references to the main plant item and drawings to which it is related.

## Sketches

### *Medium component sketch*

This serves to show in detail a medium-sized component of a plant section or item which does not require a full-sized drawing to portray it satisfactorily. It should include references to the main plant item and related drawings.

### *Composite data sketch*

This comprises a data sheet relating to a specific range of like plant or components in use, for easy comparison and assessment. It can be made to include a wide range of components and may be used to extend standardisation; in this way it may become the standard record, thereby assisting the maintenance administration functions.

### *Minor detail sketch*

This serves to portray minor details of plant items. It should include references to the main plant item and drawings to which it is related.

## METHODS OF PROJECTION

The methods of projection employed are referred to as:

*First Angle Projection.* In this case, each view is so placed that it represents the side of the object remote from it in the adjacent view.

*Third Angle Projection.* In this case, each view is so placed that it represents the side of the object near to it in the adjacent view.

Here again, it is advisable to adopt and enforce a standard. First angle projection is recommended for normal use.

## DIMENSIONING

A common source of annoyance to technical staff and shop personnel is to find that overall or composite dimensions do not stand out clearly, and may even have to be computed from a series of subsidiary dimensions.

## LIMITS, FITS AND TOLERANCES

It is surprising how vaguely this 'household' phrase in the world of precision engineering can sometimes be interpreted in heavy industry.

*Limit* refers in a general way to a particular limit system, or the extreme dimensions prescribed in order to allow for the various qualities of fits.

*Fit* refers to the type of fit, using the following classifications:

*Interference Fit*, where mating parts have to be assembled by effort or pressure; classified again into push fits, driving fits, and force fits.

*Clearance Fit*, where there is a definite clearance between the mating parts over the whole range of prescribed limits. These are again divided into various classes of running fits.

*Tolerance* refers to the amount of difference from a required dimension laid down, in order that unavoidable imperfections in workmanship can be tolerated.

*Unilateral Tolerance* is where the tolerance range is wholly above (positive), or wholly below (negative) the nominal dimension:

$$50\text{ mm } \begin{smallmatrix}+0{\cdot}050\\+0{\cdot}025\end{smallmatrix} \qquad 50\text{ mm } \begin{smallmatrix}-0{\cdot}025\\-0{\cdot}050\end{smallmatrix}$$

(refer to BS 308 (Revised)—Engineering Drawing Practice).

*Bilateral Tolerance* is where the tolerance range is above and below (positive and negative) the nominal dimension, and not necessarily equal above and below:

$$50\text{ mm } \begin{smallmatrix}+0{\cdot}02\\-0{\cdot}02\end{smallmatrix} \qquad 50\text{ mm } \begin{smallmatrix}+0{\cdot}03\\-0{\cdot}01\end{smallmatrix}$$

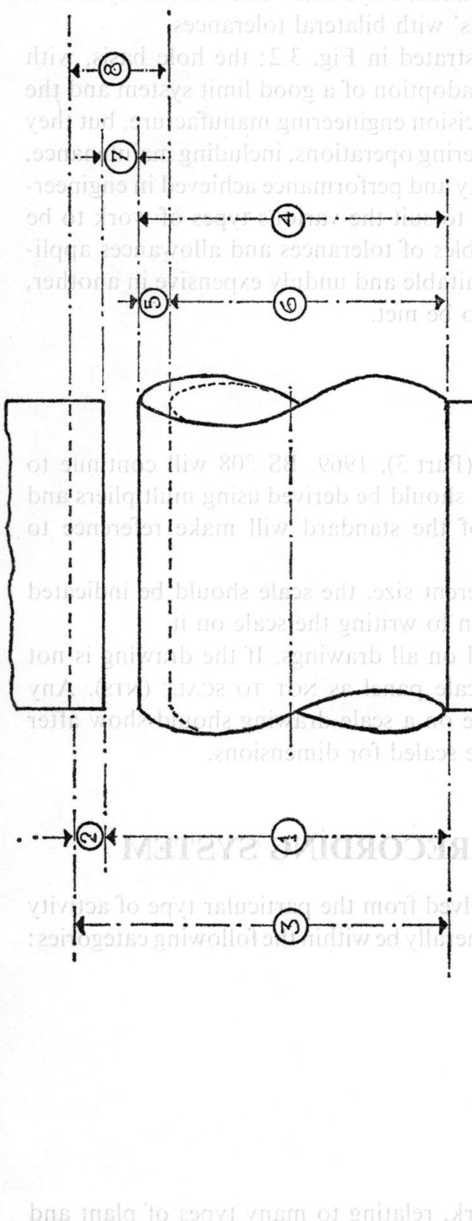


TAKING HOLE BASIS WITH UNILATERAL TOLERANCES:

| | | |
|---|---|---|
| 1 | NOMINAL DIAMETER | BASIC OR DESIGN SIZE OF SHAFT. |
| 2 | TOLERANCE | UNILATERAL (+) ON HOLE. |
| 3 | LIMIT (UPPER) | OF HOLE SIZE |
| 4 | SHAFT DIAMETER | DESIGN SIZE OF SHAFT (RELATED TO (1) ACCORDING TO CLASS OF FIT REQUIRED.) |
| 5 | TOLERANCE | UNILATERAL (−) ON SHAFT. |
| 6 | LIMIT (LOWER) | OF SHAFT SIZE |
| 7 | FIT OBTAINED | MINIMUM CLEARANCE FOR CLEARANCE FIT (+)<br>MAXIMUM INTERFERENCE FOR INTERFERENCE FIT (−) |
| 8 | FIT OBTAINED | MAXIMUM CLEARANCE FOR CLEARANCE FIT (+)<br>MINIMUM INTERFERENCE FOR INTERFERENCE FIT (−) |

FIG. 3.2. Illustration of terms—limits, fits and tolerances.

*Hole Basis* is a system of limits in which the hole is given fixed tolerances relative to a nominal diameter, and the shaft tolerances are varied to obtain the class of fit required.

*Shaft Basis* is a system of limits where the shaft is given fixed tolerances relative to a nominal diameter, and the hole tolerances are varied to obtain the class of fit required.

The classification of fits is defined in several standard systems. The Newall system is widely used in this country, and uses the 'hole basis' with bilateral tolerances.

The terms 'limits', 'fits' and 'tolerances' are illustrated in Fig. 3.2; the hole basis, with unilateral tolerances, is shown in the diagram. The adoption of a good limit system and the use of these terms are prerequisites of successful precision engineering manufacture, but they are no less important and advantageous in all engineering operations, including maintenance, and have a direct bearing on the standard of reliability and performance achieved in engineering practice. The choice of a system must be made to suit the various types of work to be undertaken, and should be made from standard tables of tolerances and allowances applicable. What is advisable in one instance can be unsuitable and unduly expensive in another, and practice should be adapted to the conditions to be met.

## SCALES

The British Standard for metric scales is BS 1347 (Part 3), 1969. BS 308 will continue to recommend that the scales for engineering drawings should be derived using multipliers and divisors of two, five, and ten, and the revision of the standard will make reference to BS 1347, Part 3.

When drawings have to be reproduced to a different size, the scale should be indicated by a scaled strip plotted on the drawing, in addition to writing the scale on it.

It is important that the scale be clearly indicated on all drawings. If the drawing is not drawn to scale, this should be indicated on the scale panel as NOT TO SCALE (NTS). Any important dimensions which are drawn out of scale on a scale drawing should show after them the abbreviation NTS. Drawings should not be scaled for dimensions.

## DRAWING NUMBERING AND RECORDING SYSTEM

The numbering and recording system is usually evolved from the particular type of activity and work engaged in. Drawings and sketches will generally be within the following categories:

*Internal origin:*

Full size drawings.
Sketches.

*External origin:*

Manufacturer's full size prints.
Manufacturer's sketches.

Where there is a great range and variety of work, relating to many types of plant and equipment, it will be found that simple sequential numbering and filing in each of the above

| DRAWING OR SKETCH NUMBER | DATE | REVISIONS (DATE) | | | | TITLE | CARD INDEX REF. | | REMARKS |
|---|---|---|---|---|---|---|---|---|---|
| | | A | B | C | D | | ACTIVITY | CATEGORY | |
| T 201 | | | | | | | | | |
| T 202 | | | | | | | | | |
| T 203 | | | | | | | | | |

FOR INTERNAL FULL SIZE DRAWING REGISTER.
FOR INTERNAL SINGLE & DOUBLE FOOLSCAP SKETCH REGISTER.

| DRAWING OR SKETCH NUMBER | DATE | REVISIONS (DATE) | | | | TITLE | CARD INDEX REF. | | FIRM | SUPPLIER'S REF. NO. |
|---|---|---|---|---|---|---|---|---|---|---|
| | | A | B | C | D | | ACTIVITY | CATEGORY | | |
| EP 201 | | | | | | | | | | |
| EP 202 | | | | | | | | | | |
| EP 203. | | | | | | | | | | |

FOR EXTERNAL MANUFACTURERS' FULL SIZE PRINT REGISTER:
FOR EXTERNAL MANUFACTURERS' SKETCH REGISTER.

FIG. 3.3. Typical pages from drawing registers (loose-leaf binders).

categories, with a suitably designed card index system to segregate and locate any drawing or sketch required, will prove suitable. Flat filing of drawings is recommended. This method enables the storage drawers to be filled each to a known and constant capacity in turn; it uses the storage space to the best advantage, and prevents the condition in which some drawers are only partly utilised, while others tend to be overloaded. The standard cabinets available will comfortably house 200 drawings or sketches in each drawer.

### The recording system

#### *Registering of drawings and sketches*

To ensure the controlled allocation of sequential numbers to drawings and sketches, it is necessary to maintain registering books for each category of drawings and sketches. Typical pages from such registers are shown in Fig. 3.3.

#### *The card index system and key*

This should be designed to give clear indication of the Unit location and plant category, together with rapid identification and location of each drawing within the filing cabinets. It is first advisable to establish a key which is appropriate to the organisation and activities it concerns. This should be clearly displayed above the card index cabinets. Its function is to define code numbers identifying the Unit and the plant category. A typical key is shown in Fig. 3.4.

The card index system should be designed to use standard proprietary cabinets and card sizes, the card printing being arranged to suit the lay-out and detail desired. It is not necessary to use special signal cards as suitable plastic strips with strong gummed linen bases are available, which can be cut to length and fixed in any position across the tops of the cards. The appropriate code numbers can be inserted into these strips. It is well to consult makers before deciding on the size of card and types of equipment to be employed. A suitable printing lay-out for the card is shown in Fig. 3.5.

## REPRODUCTION OF DRAWINGS

Reproduction of drawings may take the form of a straight reproducing process, or an enlarging or reducing process. A very wide selection of proprietary equipment is available for these purposes, and again the makers should be consulted.

## ISSUE OF PRINTS AND RECORDING OF ISSUES

Misunderstandings, errors, and confusion can easily arise if care is not taken in the issue of prints and the recording of issues and loans. Each print issued should carry the stamp of the issuing authority, and the date of issue. When appropriate, it should be prominently marked PRELIMINARY; NOT TO BE WORKED TO; NOT TO BE SCALED; FOR ESTIMATING PURPOSES ONLY; DIAGRAMMATIC ONLY; TENDERING PRINT ONLY; SUBJECT TO AMENDMENT; SUPERSEDING PREVIOUS ISSUES; FINAL. It is convenient to employ a set of bold rubber stamps for this purpose.

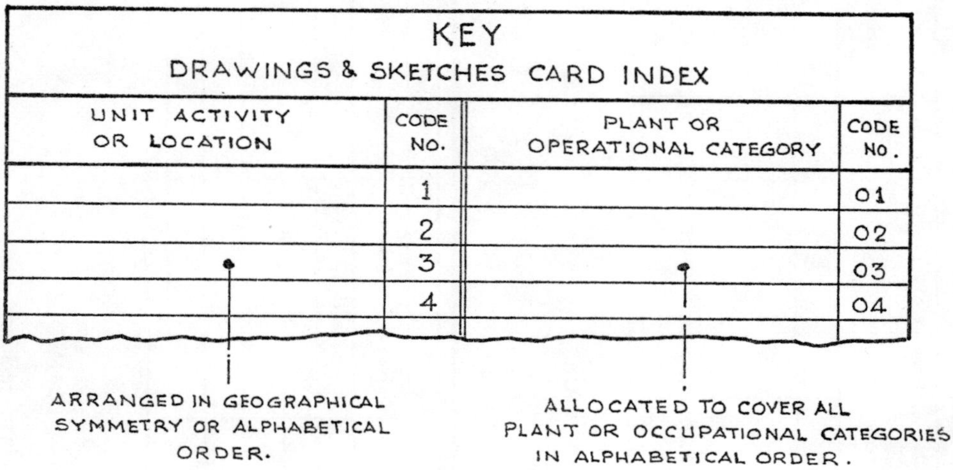


FIG. 3.4. Key to drawing office card index.

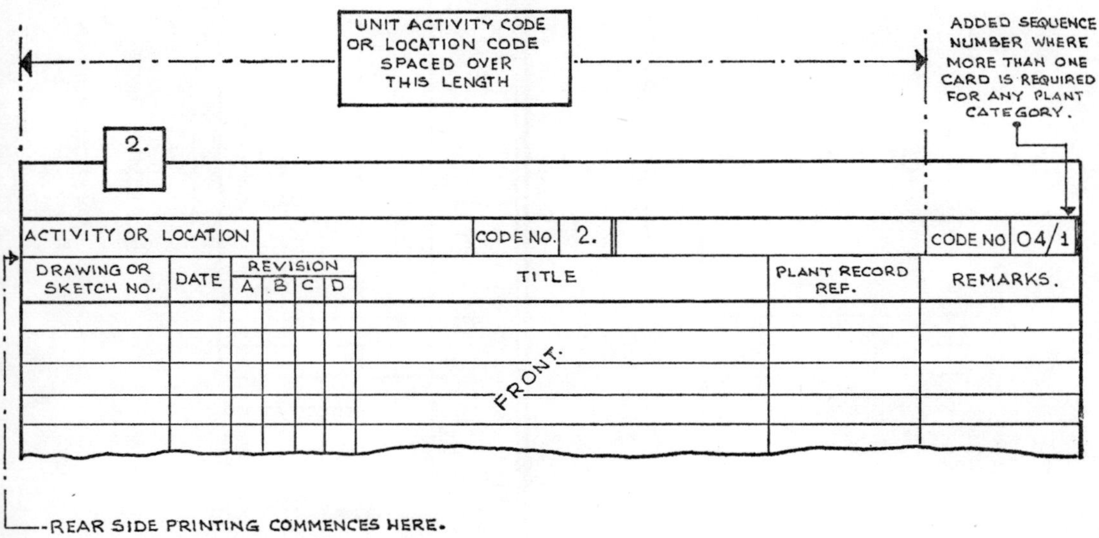


FIG. 3.5. The drawing index card.

| CARD INDEX REF | DRAWING OR SKETCH No. | REVISION | | | | DATE OF ISSUE | NO. OF PRINTS ISSUED | ADDRESSED TO | ISSUE NOTE | | SUPERCEDED OR WITHDRAWN. | DATE OF RE-ISSUE | INITIALS | |
|---|---|---|---|---|---|---|---|---|---|---|---|---|---|---|
| | | A | B | C | D | | | | SERIAL No. | RECEIPT CONFIRMED (DATE) | | | RECORD CLERK | AUTHORISED |
| | | | | | | | | | | | | | | |
| | | | | | | | | | | | | | | |
| | | | | | | | | | | | | | | |
| | | | | | | | | | | | | | | |
| | | | | | | | | | | | | | | |

SPECIMEN PAGE IN REGISTER OF PRINT ISSUES.

| CARD INDEX REF. | DRAWING OR SKETCH No. | REVISION | | | | DATE LOANED | LOANED TO | ISSUE NOTE | | RETURNED (DATE) | SIGNED |
|---|---|---|---|---|---|---|---|---|---|---|---|
| | | A | B | C | D | | | SERIAL No. | RECEIPT CONFIRMED (DATE) | | RECORDS CLERK. |
| | | | | | | | | | | | |
| | | | | | | | | | | | |
| | | | | | | | | | | | |
| | | | | | | | | | | | |
| | | | | | | | | | | | |

SPECIMEN PAGE IN REGISTER OF PRINTS LOANED.

FIG. 3.6. Specimen pages from print issues and loans registers.

The recording of issues and loans should be such that:

The destinations of all prints issued for retention or on loan, and the date of issue, are known, together with any specific instructions given with the issue.

Receipt of prints, and any instructions accompanying them, are acknowledged by the addressee.

Prints on loan, but not returned can be readily recalled and their return recorded.

The destinations of all subsequent superseding issues, the dates of issue, and the specific instructions with issue, are recorded.

Receipt of superseding issues and instructions accompanying them are acknowledged by the addressee.

All print issues are correlated with the drawing office card index system.

All recording is the duty of the drawing office records clerk, issues being countersigned by the chief draughtsman or other senior person to whom he has delegated this authority.

Clear-cut instructions can be quickly given to all concerned, and all-round liaison maintained between the drawing office and related functions. This is important, as physical errors and delays in the workshops or on site can be very costly; every effort must be made to minimise them.

The method of recording print issues and loans should form a part of the main drawing record system, which in itself establishes the necessary segregation, identification, and location related to plant, work in hand, and drawing office references. The necessary records and control can be realised by the use of the following:

| | |
|---|---|
| Register of Print Issues | see Fig. 3.6 (part). |
| Register of Prints Loaned | see Fig. 3.6 (part). |
| Print Issue and Loan Slip | see Fig. 3.7. |

A periodic check of the Register of Prints Loaned will enable such prints to be recalled, so ensuring that a cumulative depletion of the drawing office files and consequent inconvenience and disorganisation is avoided.

## STAFFING

In the consideration of staffing, it is appropriate to revert to the functions to be performed (summarised in the early part of this chapter) and to define these in broader terms, setting against them a typical staff allocation of duties. This is shown in the functional establishment chart, Fig. 3.8. The actual numbers of staff in the categories sub-ordinate to the chief draughtsman will depend upon the general administrative set-up of which the drawing office is a part, and the distribution of the functions as a whole; for example, the assessment and analysis of requirements, calculations, examination of tenders, and/or preparation of orders for the purchasing department may be allocated to the staff in the Group's engineering office; typing services may be centred in a typing pool; filing services in a central registry,

| CODE NO. PRINT ISSUE & LOAN SLIP | | | SERIAL NO. | PI / |
|---|---|---|---|---|
| | | | DATE | |
| FROM | | TO | | |
| GROUP DRAWING OFFICE | | NAME | DESIGNATION | LOCATION |
| RECORD CLERK | AUTHORISED BY. | | | |
| | | | | |

WE ENCLOSE HEREWITH THE UNDERNOTED PRINTS:

A – FOR YOUR USE AND RETENTION.

B – SUPERSEDING PREVIOUS ISSUES WHICH PLEASE DESTROY.

C – ON LOAN. TO BE RETURNED AS SOON AS POSSIBLE

* DELETE AS NECESSARY.

| NO. OF PRINTS | DRAWING OR SKETCH NUMBER | REVISION A | B | C | D | TITLE |
|---|---|---|---|---|---|---|
| | | | | | | |
| | | | | | | |
| | | | | | | |
| | | | | | | |
| | | | | | | |
| | | | | | | |
| | | | | | | |

REMARKS:

TEAR OFF

- - - - - - - - - - - - - - - - - - - - - - - - - - - -

| TO: RECORDS CLERK GROUP DRAWING OFFICE | SERIAL NO. | PI / |
|---|---|---|
| | DATE | |

THIS CONFIRMS RECEIPT OF THE PRINTS ON THE ISSUE AND LOAN SLIP FROM WHICH THIS HAS BEEN DETACHED. EFFECT WILL BE GIVEN TO THE INSTRUCTIONS THEREON.

SIGNED ........................

DATE ................ 19........

FIG. 3.7. Print issue and loan slip.

| FUNCTIONS. | DRAUGHTSMEN | | | | | TRACER | ESTIMATOR | RECORDS CLERK. | TYPIST | REMARKS |
|---|---|---|---|---|---|---|---|---|---|---|
| | CHIEF | SENIORS | GENERAL | JUNIORS | TRAINEES | | | | | |
| OVERALL SUPERVISION | ● | | | | | | | | | GROUP DIRECTION |
| DESIGN: | | | | | | | | | | |
| ANALYSIS OF REQUIREMENTS & CALCULATIONS. | ● | ● | | | | | | | | |
| LAY-OUTS & GENERAL ARRANGEMENTS. | | ● | ● | | | | | | | |
| DETAILS AND COMPONENTS. | | | ● | ● | ● | | | | | |
| CHECKING: | ● | ● | | | | | | | | |
| TRACING: | | | | | | ● | | | | |
| TAKING-OFF PRINTS: | | | | | ● | | | | | |
| RECORDING, FILING, MAILING: | | | | | | | | ● | | |
| PREPARING: | | | | | | | | | | ROUTE THROUGH GROUP ENG. OFFICE |
| SPECIFICATIONS & ENQUIRIES. | | ● | ● | | | | | | | |
| ESTIMATES. | | | | | | | ● | | | |
| CHECKING: | ● | | | | | | | | | |
| TENDERS: | | | | | | | | | | RECEIVE FROM AND RETURN TO GROUP ENGINEER^G. OFFICE |
| CHECK AGAINST ENQUIRY & ESTIMATE | ⊘ | ⊘ | | | | | ⊘ | | | |
| TECHNICAL AND FINANCIAL APPRAISAL | ⊘ | ⊘ | | | | | ⊘ | | | |
| ORDERS AND CONTRACTS: | | | | | | | | | | ON INSTRUCTIONS OF GROUP ENG^G. OFFICE WITH RETURNED DATA |
| PREPARATION FOR PURCHASING DEPT. | | ⊘ | | | | | | | | |
| PREPARATION FOR CENTRAL WORKSHOPS. | | | ⊘ | | | | | | | |
| CHECKING: | ⊘ | | | | | | | | | |
| RECORDS RELATING TO: DRAWINGS AND SKETCHES. ORDERS AND DELIVERY OF PLANT & EQUIPMENT. STANDARD CONDITIONS OF CONTRACT. INTERNAL STANDARDS/DIRECTIVES. BRITISH STANDARD SPECIFICATIONS. CATALOGUES & TECHNICAL LITERATURE. LIAISON WITH PLANT RECORDS & INVENTORY. | | | | | | | | ● | | |
| TYPING, CORRESPONDENCE & FILING. | | | | | | | | | ⊘ | POOL MAY COVER |
| INTERNAL & EXTERNAL TECHNICAL LIAISON | ⊘ | | | | | | | | | GROUP ENG^G. MAY COVER. |

● — COMMON DRAWING OFFICE FUNCTIONS.
⊘ — FUNCTIONS THAT MAY BE OUTWITH DRAWING OFFICE.

FIG. 3.8. Group drawing office functional establishment.

and mail may be routed through a central registry and dispatch office. The numbers will also depend on the type, variety, and volume of work to be handled under projects, developments, operations, and maintenance.

## ADMINISTRATIVE CONTROL

The administrative control of the drawing office is exercised by the Group's engineering establishment through the chief draughtsman, who is responsible for overall supervision and control of the day-to-day operation of the drawing office. In principle, all instructions regarding work and priorities should originate from the Group's engineering office. At the same time, a degree of flexibility is essential in certain directions; for instance, having noted that all drawing office functions and services are advisedly centred in a single establishment, it is desirable that direct service is available to the Group's workshops management on work connected with the internal day-to-day activities of the workshops. On the other hand, it is prudent to ensure that all work in connection with the Units, external to the Group's organisation, is closely controlled through the Group's engineering office. Any departure from this principle of direction should be considered and agreed, and where disputes arise or confusion develops, the matter should be investigated promptly. If necessary, remedial measures should be taken by the head of the Group's engineering organisation. If such circumstances are dealt with as they arise, cumulative disorganisation and friction are avoided. Periodic reviews of the arrangements and organisation are also desirable, to determine whether improvements can be made and a higher level of service efficiency attained, as a result of experience.

*Chapter 4*

# THE CENTRAL WORKSHOPS
## Organisation and Management for Maintenance Engineering and Jobbing Work

## GENERAL CONSIDERATIONS

In planned industrial maintenance there are many pitfalls for the unwary. These are most numerous and the consequent penalties are most severe within the organisation and operation of the central maintenance workshops. Their management policy is reflected throughout all the maintenance activities of the Group's organisation and materially affects the contribution of the engineering service to production, safety and profitability. It is essential to recognise those factors causing differences between workshop organisation, and production control on the one hand and maintenance activities on the other.

## WORKSHOP ORGANISATION AND PRODUCTION CONTROL

The basic concepts of workshop operations and production control may be broadly summarised under the following operational categories of jobbing production, batch production, and mass production.

## JOBBING PRODUCTION

Jobbing production has been practised from the earliest days of factory systems. It is applied in factories engaged on a varied range of separate jobs, meeting individual requirements. The keynote is 'making' rather than 'manufacturing'. The diversity of small product outputs involved entails the retention of commonplace workshop methods, using standard machine tools, without any particular factory lay-out considerations other than those of the available accommodation and general convenience. It demands high skills and versatility on the part of the foremen and of the craftsmen under their control. Constant vigilance is needed to improve processes and materials of manufacture. It is the hardest of all set-ups to control properly.

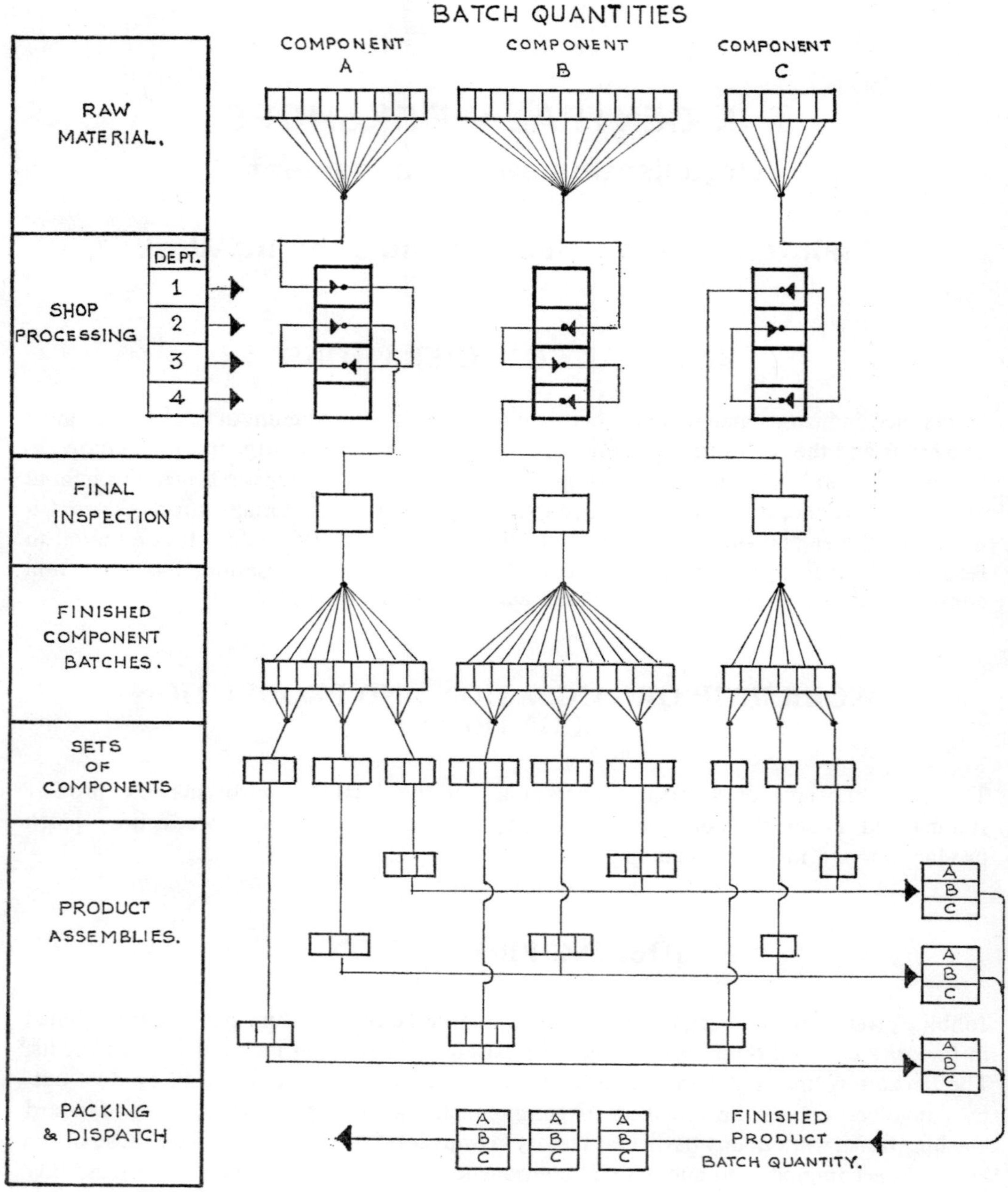


FIG. 4.1. Batch or quantity production flow diagram for a particular product.

## BATCH PRODUCTION

In this country batch production is a commonplace engineering practice. A compromise between jobbing and mass production, it allows individuality of demand with diversity of products while attaining higher outputs of the various products with economic advantage. The economic batch size of a finished product is determined by such considerations as demand, required delivery and sensible stock reserves. The components required to fulfil the production programme can be manufactured on a batch or mass production basis. This facilitates the introduction of modern production methods, including:

Controlled planning and progressing of production.
Utilising specialised machine tools, sited to afford maximum efficiency of production flow.
Close quality control of raw materials and manufacturing processes using modern inspection methods.
Routing the work on a planned basis.
Use of special jigs, tools and fixtures.
Introduction of time and motion studies.
Forward estimating and budgeting.
Application of standard costs.

Figure 4.1 illustrates batch production in elementary form, and shows how it follows a definite pattern for the planned production of a particular product. It will be apparent, however, that the scale and range of the principles and methods of batch production are influenced by the type of production involved.

## MASS PRODUCTION

Mass production infers the continuous production of standard products to closely-timed schedules, against large market demands. It involves:

Specially planned factory lay-outs.
Specialised machinery.
Expensive jigs and fixtures.
Detailed analysis and timing of operations and processes, with time, motion and method studies.
Pre-planned machine loading and production flow.
Highest standards of quality control of material and workmanship.
Full scale planning and progressing, aimed at co-ordinating the rates of component manufacture and product assembly to match the required output.
Forward estimating and budgeting.
Application of standard costs.
Specialist supervision.

It has the advantages of:

Low unit manufacturing costs.
Standardised production control.
Improved quality.
Reduction of waste of materials and labour.
Employment of supervised semi-skilled and unskilled labour with a minimum of highly skilled operatives.
Incentive bonus schemes benefiting both employees and performance.

It has the disadvantages of:

Limited flexibility of plant and facilities to meet changing conditions.
Relatively heavy cost of re-organisation and attendant lost production during product changeover.
Restricting of progress in development and design.

## WORKSHOP ORGANISATION AND CONTROL FOR SERVICE MAINTENANCE

Maintenance is a service to production, as distinct from production itself. The central workshops must be organised and operated not as an entity, but as a part of the whole undertaking. Their internal organisation, procedures, and practices must be co-ordinated with those in the engineering service organisation, of which they form a part, and with the related functions of the main organisation embracing the engineering establishment. The central workshops are the hub on which the main processes of maintenance revolve, in that they carry the burden of plant overhaul and repair on a Group basis, as distinct from site inspection and maintenance. This responsibility is augmented by other activities depending on the nature of the industrial functions being served, with the workshop facilities being able to cater for all commitments.

Service is the keynote of maintenance with production creating the service demands. Whatever form they take, however varied the problems they pose to workshop operations and control, the demands must be met in their entirety. The shops must operate to suit the industrial activities they serve and, while seeking maximum internal efficiency, their first priority is the fulfillment of the requirements of production. It is in the terms of all the varied types of work demands and the efficiency with which they are carried out, that the workshops' performance is inevitably judged. Although not readily assessable, their impact is felt in many ways, including:

The reduction of breakdowns, associated lost production and consequent profitability.
The utilisation level of production equipment and the resulting return on capital investment.
The benefits to depreciation charges due to the extended plant life, with consequent reduction of operating costs.
The deferment of capital expenditure on the replacement of plant and equipment.

A sense of the genuine priorities must be retained. It is an anomaly that for the portion of the work-load lending itself to forward planning (unlike conditions where a marketable product is involved), the time factor is not crucial; whereas in the case of the work not capable of forward planning the time factor is often critical. Once this point is appreciated it can be usefully exploited and the pitfalls of rigid, forward long-term scheduling of work-load and throughput avoided.

## BROAD ASSESSMENT OF THE NATURE OF THE WORK AND SERVICES DEMANDED

From the foregoing considerations, it is useful to assess the nature and character of the various types of work and services which make up the total demand on the shops and to set this against an estimate of their proportion of the whole. A typical assessment is shown in Fig. 4.2. This immediately emphasises that if the desired results are to be achieved, the central workshops must be organised and operated as a vital element in any undertaking. At the same time, it gives an indication of the problems likely to be encountered.

## OBJECTIVES TO BE ATTAINED

The objectives to be attained in terms of central workshop services, as part of the total engineering service organisation (which must incorporate the means of maintaining the productive equipment and services of the undertaking) can therefore be summarised as follows:

Periodic overhauls of mobile and transportable plant items.

All miscellaneous incoming repairs of plant items, components and other equipment required to sustain or service production, being dealt with on a jobbing or batch basis, according to their urgency and priority. While it is reasonable to assume that as planned maintenance becomes effective this category will become a diminishing function, it will, in the nature of things, be ever present and must be catered for.

Work concerned with the routine inspection and servicing of portable tackle and equipment under statutory regulations, to be effected on schedule to ensure safe working conditions to the insurer's satisfaction.

All necessary and economic reconditioning (for re-use) of salvaged components recovered in the course of overhaul and repair work. This work is to be carried out provided that no specialised processes beyond the shop facilities are required, and with due regard being paid to standardisation.

Small scale manufacture within the defined commitments, to be dealt with under the following headings and at competitive costings:

Repetition manufacture of non-proprietary spares and for stock replacements.

Miscellaneous jobbing and other work incidental to projects and developments not covered by external contracts.

| CATEGORY OF WORK. | | | NATURE OR CONDITIONS. | EXAMPLES. | ESTIMATED % OF TOTAL COMMITMENTS | | |
|---|---|---|---|---|---|---|---|
| MAINTENANCE. | 1 | CYCLIC OVERHAUL OF A WIDE VARIETY OF MOBILE OR TRANSPORTABLE PLANT UNITS FROM UNIT ACTIVITIES. | CONTINUOUS CYCLIC REPAIR & TURN-ROUND ON A POOL BASIS. REGULAR IN FLOW & BROAD ROUTING. IRREGULAR IN DETAIL WORK & ROUTE | PRODUCTION PLANT UNITS. SERVICE PLANT TO PRODUCTION. TRANSPORT PLANT E.G. LOCOS, LORRIES, CARS. | 40 | 80 | 100 |
| | 2 | REPAIR OF INCOMING VARIED PLANT ITEMS OUTWITH CYCLIC PROGRAMME AND INCLUDING BREAKDOWNS. | JOBBING REPAIRS. IRREGULAR IN DEMAND AND WORK DETAIL. | FAILURE OF PLANT OUTWITH POOL CATEGORY. FOR SITE REPAIRS OF COMPOSITE PLANT. | 10 | | |
| | 3 | REPAIR OF INCOMING VARIED PLANT COMPONENTS OUTWITH PROGRAMME AND INCLUDING BREAKDOWNS. | JOBBING REPAIRS. IRREGULAR IN DEMAND AND WORK DETAIL. | FOR SITE REPAIRS. FOR SERVICE STOCK | 5 | | |
| | 4 | RECONDITIONING SALVAGED COMPONENTS REPLACED DURING OVERHAUL FOR RE-USE | JOBBING REPAIR WORK. IRREGULAR IN QUANTITY AND WORK DETAIL. WIDE VARIETY. | LOCOMOTIVE WHEELS AND AXLES. | 5 | | |
| | 5 | BATCH REPAIRS OF MULTIPLE LIKE ITEMS. | BATCH REPAIR WORK. CAPABLE OF PRODUCTION LINE MAINTENANCE, EVEN FLOW | RAILWAY WAGONS. TRAILING CABLES. | 5 | | |
| | 6 | INSPECTION & SERVICING OF TACKLE & EQUIPMENT | ROUTINE WORK OF REGULAR VOLUME CORRELATED WITH STATUTORY REQUIREMENTS. | LIFTING TACKLE. BOILER FITTINGS. | 5 | | |
| | 7 | MISCELLANEOUS WORK FOR SITE REPAIRS AT UNIT ACTIVITIES.<br>LABOUR & TACKLE SERVICES ON WORK AT UNIT ACTIVITIES. | TO MEET SCHEDULED PROGRAMME OF SITE WORK.<br>SCHEDULED AND UNSCHEDULED SERVICE. | MATERIALS FOR HOLIDAY PERIOD SITE REPAIRS.<br>PLANT OVERHAULS AND REPAIRS IN SITU. | 10 | | |
| MANUFACTURE. | 8 | NON-PROPRIETARY SPARES AND RENEWALS FOR STOCK AGAINST FUTURE REQUIREMENTS. | REPETITION MANUFACTURE OF VARIED NATURE. IRREGULAR AND AS CONDITIONS PERMIT. | LOCOMOTIVE CONNECTING ROD BUSHES. | 5 | 20 | |
| | 9 | WORK INCIDENTAL TO PROJECTS AND DEVELOPMENT WORK IN HAND AND NOT COVERED BY EXTERNAL CONTRACTS. | JOBBING MANUFACTURE OF MINOR NATURE. IRREGULAR. TO MEET SITE PROGRAMME PRIORITIES | LIGHT STRUCTURAL STEELWORK AND FABRICATIONS. | $7\frac{1}{2}$ | | |
| | 10 | SUNDRY ITEMS OF NO DEFINITE CLASSIFICATION. | JOBBING MANUFACTURE OF VARIED PRIORITY. | SMALL PROTOTYPE OR EXPERIMENTAL EQUIPMENT. | $2\frac{1}{2}$ | | |
| | 11 | LABOUR & TACKLE SERVICES ON SITE DEVELOPMENTS, INSTALLATION, AND ERECTION WORK. | TO SCHEDULED PROGRAMME OR AS OCCASION ARISES. | SITE PLANT EXTENSIONS, MODIFICATIONS AND RECONSTRUCTIONS. | 5 | | |

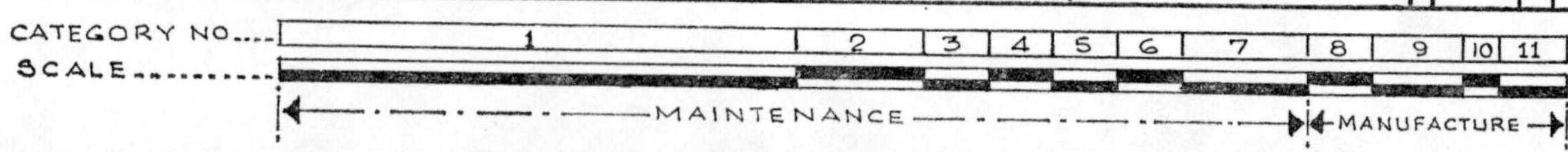


FIG. 4.2. Workshops work load assessment chart.

Services for site overhaul, repair or minor installation work at the Units.
Workshops to maintain their own equipment to required standards.
Apprentice craftsmen to be given comprehensive training in the principles and practice of their trades, and facilities for personnel to attend special centres in justifiable cases, for example, welding schools.
The achievement and maintenance of standards of workmanship consistent with economy.
All work and services to be accurately costed and charged.
All necessary records of work executed (with costings) to be maintained; also other required information, reports, and returns to be issued to the agreed recipients.
All practices and procedures within the workshops to be satisfactorily co-ordinated with the other functions of the general engineering organisation and the related functions of the Group's organisation as a whole.
The workshops to be directed, supervised, organised and operated so that their commitments are met satisfactorily and efficiently, and with good personnel relationships maintained.

## THE TYPE OF ORGANISATION AND CONTROL

The overhaul function of maintenance, as applying to central workshop activities, introduces more complex and varied conditions and problems than those encountered in production systems of manufacture. In practice, a product is received in a secondhand and unknown condition. It is dismantled, thereby exposing a further set of previously unknown factors and requirements. Further similar plant items and those of a different nature make up the work load. Each item produces factors and requirements differing in task and detail. Figure 4.3 gives an illustration of the overhaul process applied to a single mobile or transportable plant item. The disarray arising from simultaneous handling of like and unlike categories of plant, as compared with the readily developed pattern of manufacturing production will be apparent (*see* Fig. 4.1). The position is further complicated by miscellaneous jobbing and batch work and labour demands for external or site work.

The answers to the problems are not always as straight-forward as they sometimes appear. For example, it might be considered that the periodic overhaul work load, although having different requirements according to the nature of the plant items, could be planned and progressed to a rigid forward programme on a timed batch basis, the units being stripped down and any components not fit for immediate re-use being replaced by new parts, with the discarded components being collected and reconditioned on a batch recovery basis and placed in stock. In practice this overlooks certain factors. The main housing or structure of the machine itself (into which normal replacements are fitted) frequently reveals defects that preclude direct component replacement, for example worn bearing housings, worn spigots, faulty facings, fractures, and so on. Decisions have to be taken on whether to restore or replace. Again, a prerequisite to a successful batch recovery process is standardisation of both the plant item and its components. With such varied work and the changes in the unit according to the model number, it becomes evident that a rigid policy is impracticable.

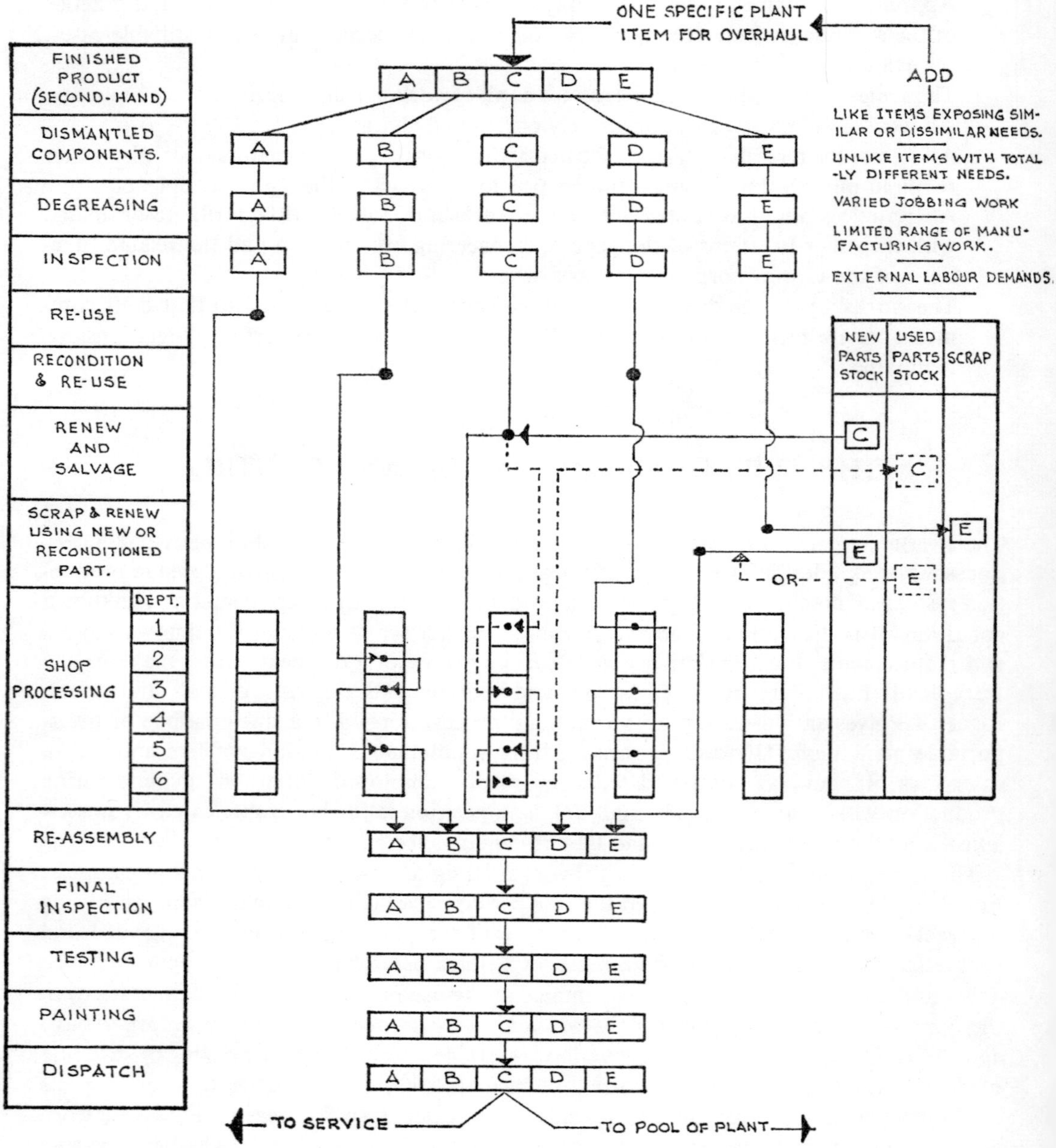


FIG. 4.3. Maintenance flow diagram illustrating overhaul process for a specific plant item.

The treatment of each plant unit on a common basis, with the overhaul processes embracing all the required and economic reconditioning of parts as a standard operation, is correct in principle. However, a certain amount of after-recovery work on components is worth while with specific types of plant.

The factors to be borne in mind are therefore:

Service to production, in all its forms and demands, is the first consideration.

Efficiency of service is reflected and judged by the Unit's production. While the aim is to maintain as high a standard of workshop efficiency as possible, the ultimate test is production output.

The labour force and facilities must be continually regulated and deployed in the interests of the Group as a whole; the true priorities are maintained by a broad measure of control from the Group's engineering office.

The internal procedures and practices must be effectively correlated with those in related functions within the Group's organisation.

The dominant feature of the work-load—the periodic overhauls—entail forward planning, but must not be regarded as truly repetitive, or essentially to be carried out to forward, rigid, long-term schedules. Rather they should be dovetailed into the overall activities in such a way that the required throughput of plant is attained on average over the year; this flexible characteristic is used to avoid undesirable repercussions in other directions. This feature eases adverse effects on the workshops' overall performance due to delay in availability of materials or labour, and aids assessment of the genuine priorities on a day-to-day basis.

It is essential that close on-the-spot direction and supervision is exercised on the shop floor, giving full scope for experienced judgement and initiative by the section foremen and chargehands.

## SITE AND LOCATION

It is desirable that the workshops are located as an integral part of the engineering establishment at the Group's headquarters. This may not always be possible, particularly when reorganisation of existing facilities is the plan. At the least, close day-to-day contact must be maintained between the Group's engineering office and the central workshops. It is not proposed to detail here the various considerations bearing upon the choice of the site and location, but the following points are a brief guide:

Geographical convenience in relation to the Group's headquarters and the Units.

Geological suitability of site.

Road and rail access, and communications generally.

Availability of power and water supplies.

Availability of labour, housing, and amenities.

Possible future local planning developments, or diminishment of the district's suitability due to transfer of industry, redundancy, and developments elsewhere.

## ASSESSMENT AND ANALYSIS OF THE WORK LOAD

Before considering a workshop's layout and facilities, it is necessary to assess and analyse the work load, and the service facilities and labour required. The plant record and inventory can make a useful contribution with regard to the dominant function of periodic plant overhaul and repairs. The other functions of jobbing repairs, component reconditioning, batch repair work, tackle servicing, service manufacturing, structural steelwork fabrication, smithy forging and site services, are less readily assessed.

### Periodic overhauls

Figures 4.4 and 4.5 illustrate the compilation of data, the contributory considerations and the summarised conclusions which, taking into account all the categories of plant included, give a fair assessment of the periodic work-load in terms of the nature of work, throughput, manpower, trades and floor space required.

### Jobbing repair and overhaul work

To avoid encroachment on the floor sections allocated to periodic plant overhaul and the diversion of the labour employed in these sections, it is advisable to make full allowance for the other service facilities. This mitigates the impact of emergencies and irregularities upon the periodic programme, covers plant items not included in the periodic programme, and segregates miscellaneous component reconditioning from the periodic operations. Referring to Fig. 4.2, showing the estimated analysis of the total demand on the shops and proportionate claims under each heading, it is seen that work under these headings was estimated at 25%, 12½% and 12½% respectively of that for periodic overhauls. Given this basis, the following additional assessment might be made, covering the work in these categories:

| CYCLIC OVERHAUL WORK | | | | | | | | | | | | | | | | | |
|---|---|---|---|---|---|---|---|---|---|---|---|---|---|---|---|---|---|
| % of Total Shops Commit-ments | Manpower in Trade Categories (Summary)* | | | | | | | | | | | | | | | | Estimated Floor Space Area |
| | I | FG | FD | LF | M | PE | B | W | S | U | O | C | P | EW | EG | Total | |
| 40% | | | | | | | | | | | | | | | | | |
| On Incoming Miscellaneous Plant Items Excluded from Programme | | | | | | | | | | | | | | | | | |
| 10% | | | | | | | | | | | | | | | | | |
| On Incoming Miscellaneous Plant Components | | | | | | | | | | | | | | | | | |
| 5% | | | | | | | | | | | | | | | | | |
| On Reconditioning of Salvaged Components | | | | | | | | | | | | | | | | | |
| 5% | | | | | | | | | | | | | | | | | |

* For key see Fig. 4.4.

| ITEM No. | PLANT | | | | GROUP/UNIT HOLDING | | | | | | FOR ASSESS-MENT PURPOSES | PERIOD BETWEEN OVERHAULS. (YEARS) | THROUGHPUT IN UNITS/ANNUM OF 48 WEEKS. |
|---|---|---|---|---|---|---|---|---|---|---|---|---|---|
| | CATEGORY | MAKER | TYPE OR MODEL | OPS. CARD REF. | IN SERVICE | FLOATING POOL | | EMERGENCY STOCK. | TOTAL | PLUS OR MINUS NEXT 5 YEARS. | | | |
| | | | | | | UNDER REPAIR | HELD AND IN TRANSIT. | | | | | | |
| 1 | COMPRESSOR | XY & CO. | M15 | 1/CPR | 96 | 3 | 3 | - | 102 | - | 96 | 2 | 48 |
| | | | | | | | | | | | | | |
| | | | | | | | | | | | | | |

| MAN-HOURS PER UNIT | | | | | | | | | | | | | | | | ANNUAL MAN-HOURS TOTAL | UNIT THROUGHPUT RATE (AVERAGE) DAYS. | PERSONNEL | | | | | | | | | | | | | | | | SHOP FLOOR AREA REQ[D] SQ. FT |
|---|---|---|---|---|---|---|---|---|---|---|---|---|---|---|---|---|---|---|---|---|---|---|---|---|---|---|---|---|---|---|---|---|---|---|
| TRADE CATEGORIES. | | | | | | | | | | | | | | | TOTAL | | | TRADE CATEGORIES. | | | | | | | | | | | | | | | TOTAL | |
| I | FG | FD | LF | M | PE | B | W | S | U | O | C | P | EW | EG | | | | I | FG | FD | LF | M | PE | B | W | S | U | O | C | P | EW | EG | | |
| | 60 | | | 20 | | | | | 20 | | | | | | 100 | 4800 | 5 | | 1·5 | | | 0·5 | | | | | 0·5 | | | | | | 2·5 | 190 |
| | | | | | | | | | | | | | | | | | | | | | | | | | | | | | | | | | | |
| | | | | | | | | | | | | | | | | TOTALS. | | | | | | | | | | | | | | | | | | |

| TRADE CATEGORY KEY | | | | | | | | | |
|---|---|---|---|---|---|---|---|---|---|
| I | INSPECTOR | LF | FITTER'S LABOURER | B | BURNER | U | UNSKILLED (GENERAL) | P | PAINTERS. |
| FG | FITTER (GENERAL) | M | MACHINIST. | W | WELDER | O | OILER/GREASER | EW | ELECTRIC WINDER. |
| FD | FITTER (DIESEL) | PE | PLATER/ERECTOR | S | SEMI-SKILLED | C | CLEANERS. | EG | ELECTRICIAN (GENERAL) |

FIG. 4.4. Periodic repairs and overhauls—mobile and transportable plant units. Build-up of shop load in terms of plant category, throughput, manpower, trades and floor space.

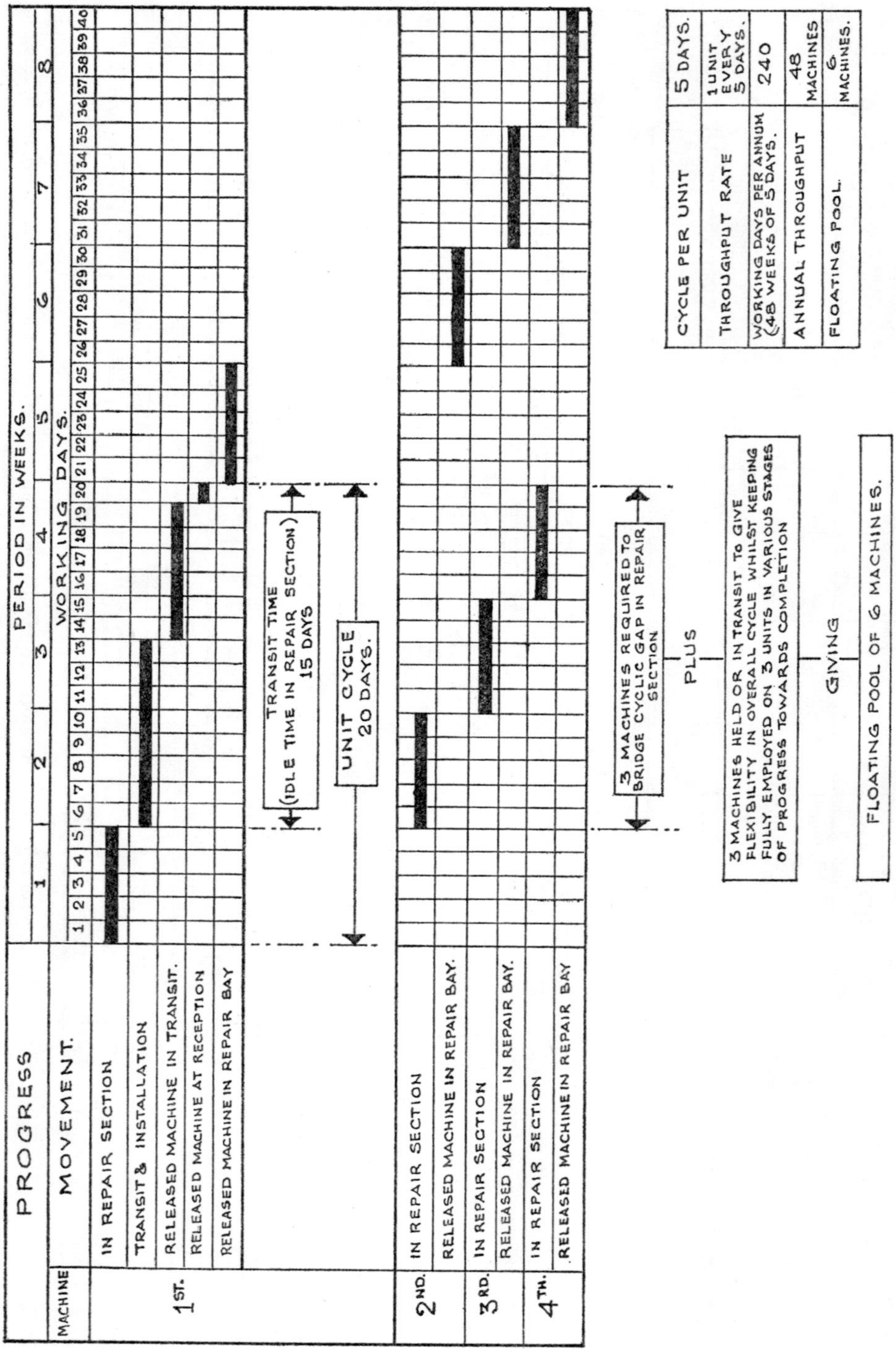


FIG. 4.5. Periodic chart for a typical plant unit category across its shop floor section.

Floor space allocated under these headings must be related to the size and type of overhaul work to be dealt with and the number of personnel so engaged. All three categories of maintenance work are irregular in nature, but only the first two are required to deal with breakdowns and emergencies, while the third is routine servicing. There is thus flexibility in coping with urgent demands, and labour is kept usefully employed in all three sections.

## Other maintenance work

### *Batch repair work, routine servicing of tackle and equipment, miscellaneous work for site repairs*

Again, it is advisable to avoid encroachment into the allocated floor sections and labour. A study of the volume and nature of the work under these headings should be made so that an assessment of floor space requirements, shop and sectional locations, and personnel requirements can be made; also of additional servicing facilities needed. These assessments should be related to the percentage claims, under their headings, of the total demand on the shops (Fig. 4.2), to maintain the overall balance of the work-load.

### *Labour and tackle for site overhaul and repair work*

The provision of labour and tackle for the repair and overhaul of fixed plant and equipment at the Units must be included. This work will usually fall within the following categories:

Work which can only be carried out at scheduled holiday periods, and for which forward programmes can be drawn up.

Work which can be carried out at week-ends or during overtime.

Work which can be carried out during normal working periods, because of the availability of standby plant.

Work which calls for immediate attention to sustain production.

The assessments for the periodic overhaul and repair work passing through the shops are based on a 48-week year, which takes into account the effects of holidays, but does not permit the diversion of labour to other purposes. Conversely, there is considerable flexibility in other portions of the work-load, permitting some diversion of labour. Since the main burden of site overhauls and repairs will fall under the first category above, the policy for site repairs and overhauls may be summarised as follows:

Arrange for the central workshops personnel to stagger their holidays so that adequate labour and services are available during the scheduled holiday periods at the Units.

Arrange for engineering personnel at the Units similarly to adjust their holidays to give the fullest aid with the holiday site repair and overhaul programmes.

Prepare (in advance) phased programmes for holiday work periods, clarifying labour, service and material requirements.

Meet non-holiday period demands for site overhaul and repair work by employing on overtime, or diverting from shop work, labour not engaged on periodic or scheduled work in the shops. Union agreements and attitudes may require to be taken into consideration in this context.

Such arrangements give shop personnel spasmodic increases in their earnings and avoid the need to engage additional maintenance staff.

### *Workshops internal maintenance*

Apart from cleaning, and routine attention such as lubrication, the physical maintenance of equipment within the workshops can be readily covered by periodic small-scale diversion of appropriate labour.

### *Manufacture*

While manufacture is not usually thought of as a maintenance function it must be accepted as a part of the total work load on the central workshops and it makes demands upon the facilities available. Referring again to Fig. 4.2, a maintenance/manufacture ratio of 80% to 20% is shown. In the objectives already set out it is aimed to undertake small-scale manufacture within the specified limits. While the best possible service must be given, the maintenance/manufacture ratio should not be allowed to change to the detriment of the main purposes of maintenance.

A study of the volume and nature of manufacturing demands likely to arise, based on local operational experience and development trends, should be made in order than an assessment of the shop, departmental and sectional floor spaces, facilities and personnel requirements may be drawn up. This assessment should be related to the percentage claims of the total work-load, with a view to maintaining overall balance.

## ASSESSMENT OF OVERALL OPERATIONAL FACILITIES AND SERVICES REQUIRED

Having made a broad assessment of the work-load, it is now necessary to establish the operational facilities and services required to cope with it. While the foregoing considerations offer some guidance, they do not provide answers to all the questions. This requires judgement; it is also prudent to accommodate, equip and staff on a scale falling short of the assessed requirements; extension of floor space and augmentation of equipment and staff should be allowed for and can be adjusted as the work load builds up and experience is gained to highlight any deficiencies. A typical assessment is shown in Fig. 4.6.

## ACCOMMODATION AND LAY-OUT

It should be stressed at once that perfection cannot be obtained and that compromise must be made to secure the optimum overall advantage; a practical approach based on experience in the conditions applying is necessary.

It is convenient first to determine the basic functional departments of the workshops, with their manning and floor space requirements. These can be prepared in tabular form under the headings of Functional Department, Section, Sub-Section, Manning and Floor Space.

For a general approach to lay-out, these can be condensed to the accommodation headings of Section, Sub-Section and Floor Space.

<table>
<tr><th>BASIC ACTIVITY</th><th>MECHANICAL</th><th>ELECTRICAL</th><th>STEELWORK FABRICATION ERECTION</th><th>SMITHY</th></tr>
<tr><td rowspan="17">PARTICULAR SERVICE FACILITIES AND OPERATIONS</td><td colspan="2">RECEPTION & OFF-LOADING</td><td colspan="2" rowspan="2">SECTIONAL STEEL & PLATE STORAGE</td></tr>
<tr><td colspan="2">HOLDING STORAGE</td></tr>
<tr><td colspan="2">STRIPPING</td><td>COLD SAWING</td><td>HAND FORGING</td></tr>
<tr><td>DEGREASING</td><td>CLEANING</td><td rowspan="2">BURNING</td><td rowspan="2">POWER FORGING</td></tr>
<tr><td colspan="2">INSPECTION</td></tr>
<tr><td colspan="2">ASSESSMENT.</td><td>WELDING</td><td>ANNEALING</td></tr>
<tr><td colspan="2">HOLDING STORAGE.</td><td rowspan="2">PUNCHING & SHEARING</td><td rowspan="2">NORMALISING</td></tr>
<tr><td>MARKING-OFF</td><td>WINDING</td></tr>
<tr><td>MACHINING</td><td>STOVING & IMPREGNATING</td><td>RIVETTING</td><td></td></tr>
<tr><td>FITTING</td><td>VULCANISING</td><td rowspan="2">PRESSING</td><td rowspan="2"></td></tr>
<tr><td>ASSEMBLY</td><td>FITTING & ASSEMBLY.</td></tr>
<tr><td>INSPECTION</td><td>INSPECTION</td><td>BENDING AND STRAIGHTENING.</td><td></td></tr>
<tr><td colspan="2">JOINT ASSEMBLY</td><td rowspan="2">ROLLING</td><td rowspan="2"></td></tr>
<tr><td colspan="2">INSPECTION</td></tr>
<tr><td colspan="2">TESTING</td><td>SCREWING</td><td></td></tr>
<tr><td colspan="2">LOADING & DISPATCH</td><td rowspan="2">ERECTION</td><td rowspan="2"></td></tr>
<tr><td></td><td>BATTERY CHARGING.</td></tr>
<tr><td rowspan="18">GENERAL SERVICES AND FACILITIES.</td><td colspan="4">WORKS STORE WITH OIL & TACKLE ACCOMMODATION</td></tr>
<tr><td colspan="4">PATTERN STORE</td></tr>
<tr><td colspan="4">TOOL ROOM</td></tr>
<tr><td colspan="4">INTERNAL & EXTERNAL TRANSPORT.</td></tr>
<tr><td colspan="4">CRANAGE</td></tr>
<tr><td colspan="4">SCRAP AND REFUSE DISPOSAL</td></tr>
<tr><td colspan="4">COMPRESSED AIR SERVICE</td></tr>
<tr><td colspan="4">WATER SERVICE</td></tr>
<tr><td colspan="4">POWER SUPPLY</td></tr>
<tr><td colspan="4">LIGHTING</td></tr>
<tr><td colspan="4">HEATING AND VENTILATING</td></tr>
<tr><td colspan="4">PAINTING</td></tr>
<tr><td colspan="4">WASHING & CLOAK ROOMS & TOILETS</td></tr>
<tr><td colspan="4">FIRST-AID SERVICE</td></tr>
<tr><td colspan="4">TRAINING FACILITIES</td></tr>
<tr><td colspan="4">FIRE PRECAUTIONS.</td></tr>
<tr><td colspan="4">TEST HOUSE</td></tr>
<tr><td colspan="4">CYCLE AND CAR ACCOMMODATION</td></tr>
<tr><td rowspan="4">ADMINISTRATIVE</td><td colspan="4">MANAGEMENT & SUPERVISORY WITH ACCOMMODATION</td></tr>
<tr><td colspan="4">PLANNING & PROGRESSING</td></tr>
<tr><td colspan="4">TIMEKEEPING</td></tr>
<tr><td colspan="4">COSTING</td></tr>
</table>

FIG. 4.6. Chart of overall operational facilities and services.

Next, there are some basic features to be noted:

So far as practicable, work should pass through the workshops in a progressive flow-path from receipt to dispatch, with the most expeditious routing for the heaviest flow of work. Movement of labour and materials should be kept to the minimum.

There should be clear segregation of the main departments and of their sections and sub-sections.

Ample and clearly defined walkways should be determined and maintained, giving access through and across the various departments and to focal points within them. These should permit passage of internal transport as required by the activities involved.

Unless they are necessary (as for example in a welding section, or to contain noise or fumes) mid-walling or partitioning should be avoided, and an open-plan concept adopted.

Single-level layout, without stairways and steps, is to be preferred and only departed from on the grounds of space, site limitations, or over-riding economic considerations.

Sections in which noisy processes and vibration take place should be kept as remote as possible from office accommodation, test-beds, the marking-off section and the machine shops. It is preferable to keep the test department, with its sensitive equipment, away from the main workshops area.

While motor transport and locomotive sections come within the general engineering department, they call for special provisions in the form of inspection pits and handling arrangements and should be set apart in a location convenient to road and rail access.

Desirably the planning and progress office should have a general view over the shop floors with direct access to them. The foremen's offices should be so arranged that each foreman has a general view over his department. It may be helpful if the foremen's offices can be placed within view of each other.

Having regard to the nature of the work involved the machine shop should comprise non-specialised general purpose machinery, laid out on the 'group' system, each unit being fitted with its own prime mover.

The general engineering and electrical test-beds should be near to each other to aid running tests of composite plant unit assemblies.

The tool room and internal stores should be located as conveniently as possible to the operating departments.

The maximum coverage of floor space, with ample headroom and minimum of obstruction, should be achieved by overhead cranes and suitable means of rapid cross-transport between the main crane bays should be provided. Principal considerations are off-loading and storing; internal handling, within departments, to the appropriate floor section and to appropriate machines; intermediate handling between departments; and storage and dispatch loading.

The boiler house, sub-station and compressor house should be segregated from the main workshop block, but as near as permissible in order to avoid undesirable extension of cabling and distribution pipework.

Thought should be given to the possibility of future extension in any department, and its accomplishment with the minimum of interruption, disorganisation and cost.

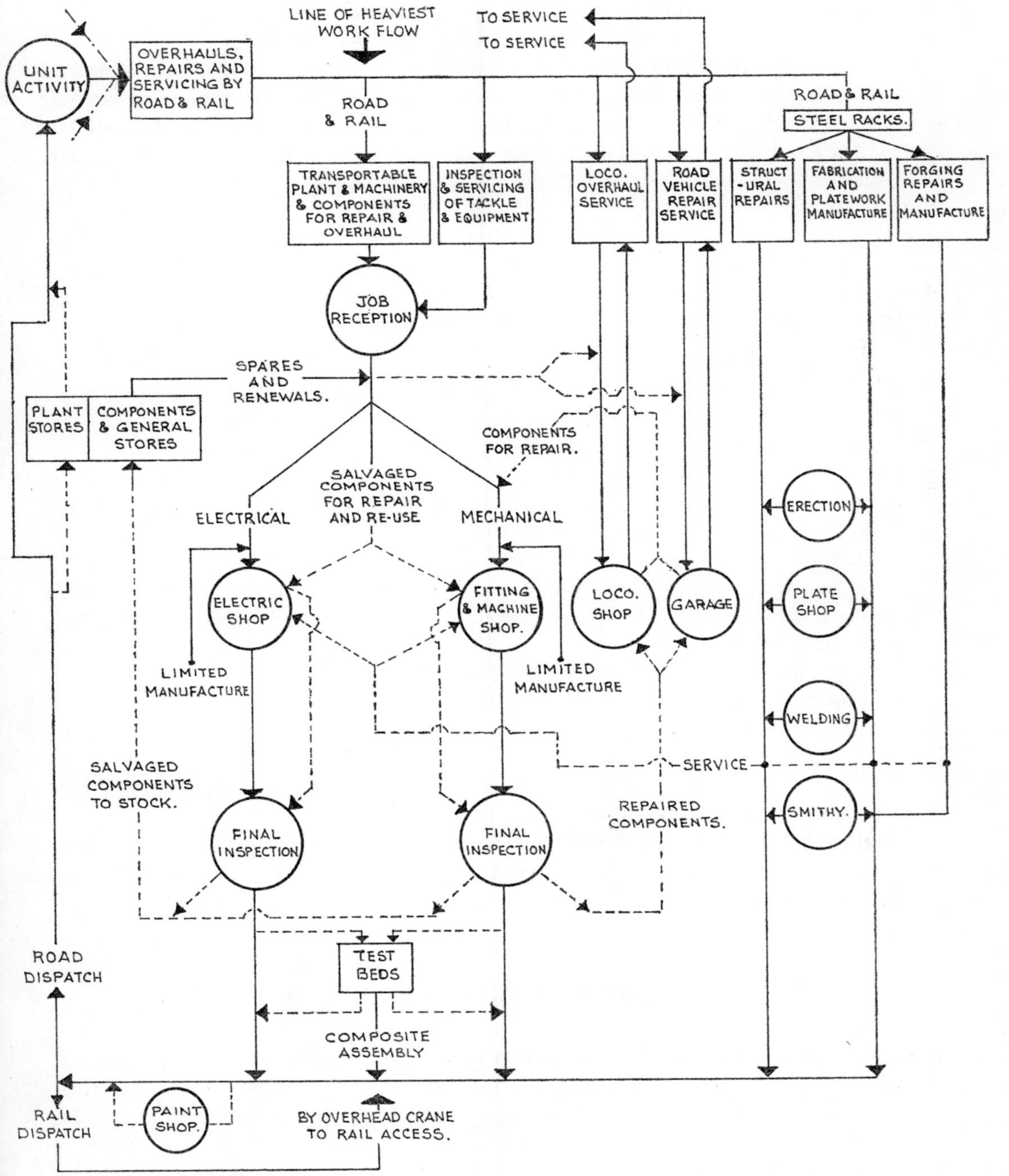


FIG. 4.7. Central workshops work flow diagram.

It is now appropriate to prepare a straight-forward work flow diagram, at the same time taking into account the various classes of work involved (*see* Fig. 4.7).

From this and preceding information, a block lay-out can be drawn out as shown in Figs. 4.8 and 4.8(a), from which detailed overall and sectional lay-outs may be prepared. It is obvious that the road and rail access, the conditions and extent of the site available, and the operations to be catered for influence the final arrangement for any undertaking. The form of building to enclose the layout is influenced by the type of processes to be undertaken.

In the case considered, on examination of the work flow diagram, the single-level bay form of construction is recommended; it lends itself to future extension by the addition of an extra bay or part-bay, or by the lengthening of existing bays. In designing the buildings, the following points should also be considered:

They must comply with legal requirements, local bye-laws, and insurance requirements.
Bay span and length is a function of the required floor space and the number of bays.
The span of each bay should be limited to a maximum accepted standard.
Glazing should be provided to give uniform natural light over the interior.
Good and well-distributed artificial lighting is essential. A combination of general overhead lighting, combined with local lighting where a higher intensity of illumination is necessary, is generally recommended.
A level concrete floor should be provided over the whole enclosed area, with a hardened finish to reduce dust and aid cleanliness. In the floor construction, provision should be made for all machinery fixing, erection, cable ducts and services to minimise after-cutting.
Fire-resistant building materials should be employed.
Adequate and well-distributed exits should be provided.
Special precautions are required where combustible and inflammable materials are used or stored.
No adverse environmental effects should be allowed; for example, consideration should be given to the extraction of process fumes.
Heating and ventilation, as factors bearing directly upon health and efficiency, should receive careful attention, regard being given to any operational processes calling for particular investigation.

## FUNCTIONAL ORGANISATION AND STAFFING

A chart of organisation, classifying the various designations and functions of the personnel, and the lines of authority and responsibility, together with complementary Group functions and authorities is given in Fig. 4.9. From this, a personnel register can be devised using proprietary designs of indexed panel frames, carried on wall brackets, rotary stands or desk stands. These give quick line-by-line reference, each line of information being carried on a replaceable slip, a facility which enables changes of personnel to be readily noted, and the records easily kept up-to-date. Tinted line inserts and distinctive signals provide clear

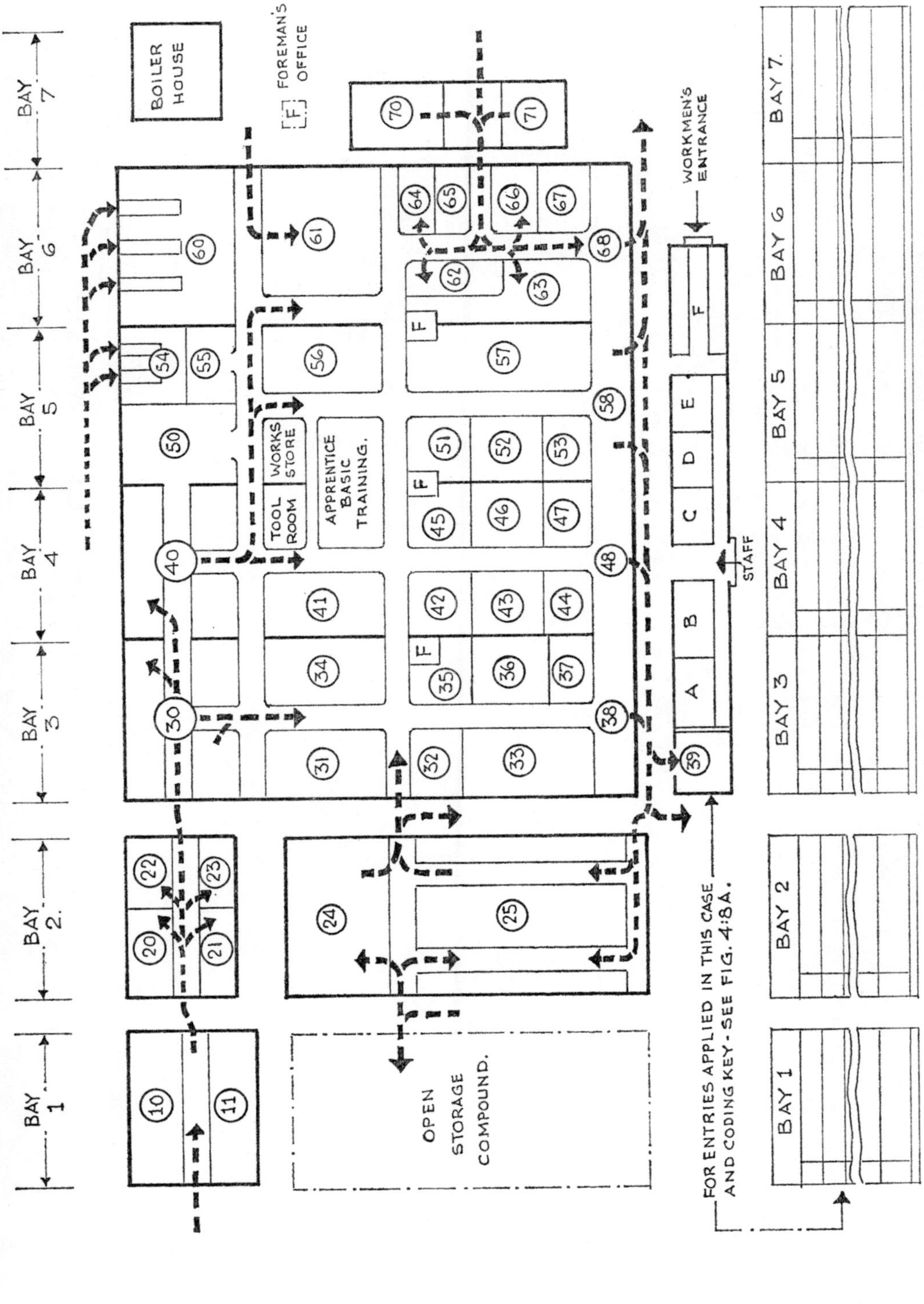


FIG. 4.8. Block lay-out of central workshops with sub-sections and operations code.

OFFICE BLOCK.

| | | | |
|---|---|---|---|
| A | MEETING ROOM. | D | COST OFFICE. |
| B | PLANNING & PROGRESS OFFICE. | E | TIME OFFICE. |
| C | MANAGER'S OFFICE. | F | CLOAK & WASH ROOMS. |

WORKSHOPS SUB-SECTIONS & OPERATIONS.

| BAY 1. | | BAY 2. | | BAY 3. | | BAY 4. | | BAY 5. | | BAY 6 | | BAY 7. | |
|---|---|---|---|---|---|---|---|---|---|---|---|---|---|
| 10 | HOLDING AREA (MECHANICAL). | 20 | STRIPPING | 30 | HOLDING AREA (ELECTRICAL) | 40 | HOLDING AREA (MECHANICAL). | 50 | JOINERS' SHOP | 60 | LOCO. SHOP | 70 | PLATE RACKS. |
| 11 | HOLDING AREA (ELECTRICAL). | 21 | DEGREASING. | 31 | WINDING | 41 | TURNING & BORING | 51 | CONVEYOR REPAIRS | 61 | SMITHY. | 71 | SECTIONAL STEEL AND BAR RACKS. |
| | | 22 | INSPECTION (MECH.) | 32 | STOVING AND IMPREGNATING | 42 | DRILLING | 52 | I.C. ENGINE REPAIRS. | 62 | SCREWING | | |
| | | 23 | INSPECTION (ELECT.) | 33 | CABLE REPAIRS | 43 | SLOTTING. MILLING AND PLANING. | 53 | GENERAL REPAIRS (MECH.) | 63 | PLATE SHOP. | | |
| | | 24 | CENTRAL STORES (LIGHT) | 34 | GENERAL REPAIRS. | 44 | TEST BED (MECHANICAL). | 54 | MOTOR GARAGE | 64 | PROFILE BURNING | | |
| | | 25 | CENTRAL STORES (PLANT) | 35 | SWITCHGEAR REPAIRS. | 45 | PUMP REPAIRS. | 55 | BATTERY & TRUCK CHARGING | 65 | METAL SAWING | | |
| | | | | 36 | TRANSFORMERS | 46 | COMPRESSOR REPAIRS. | 56 | ELECTRIC WELDING | 66 | DRILLING PUNCHING SHEARING | | |
| | | | | 37 | TEST BED (ELECTRICAL) | 47 | HAULAGE, HOIST, AND WINCH REPAIRS. | 57 | FABRICATION & ERECTION | 67 | PRESSING | | |
| | | | | 38 | DISPATCH | 48 | DISPATCH | 58 | DISPATCH. | 68 | DISPATCH | | |
| | | | | 39 | PAINT SHOP. | | | | | | | | |

FIG. 4.8(a). Key to codings in central workshops block lay-out.

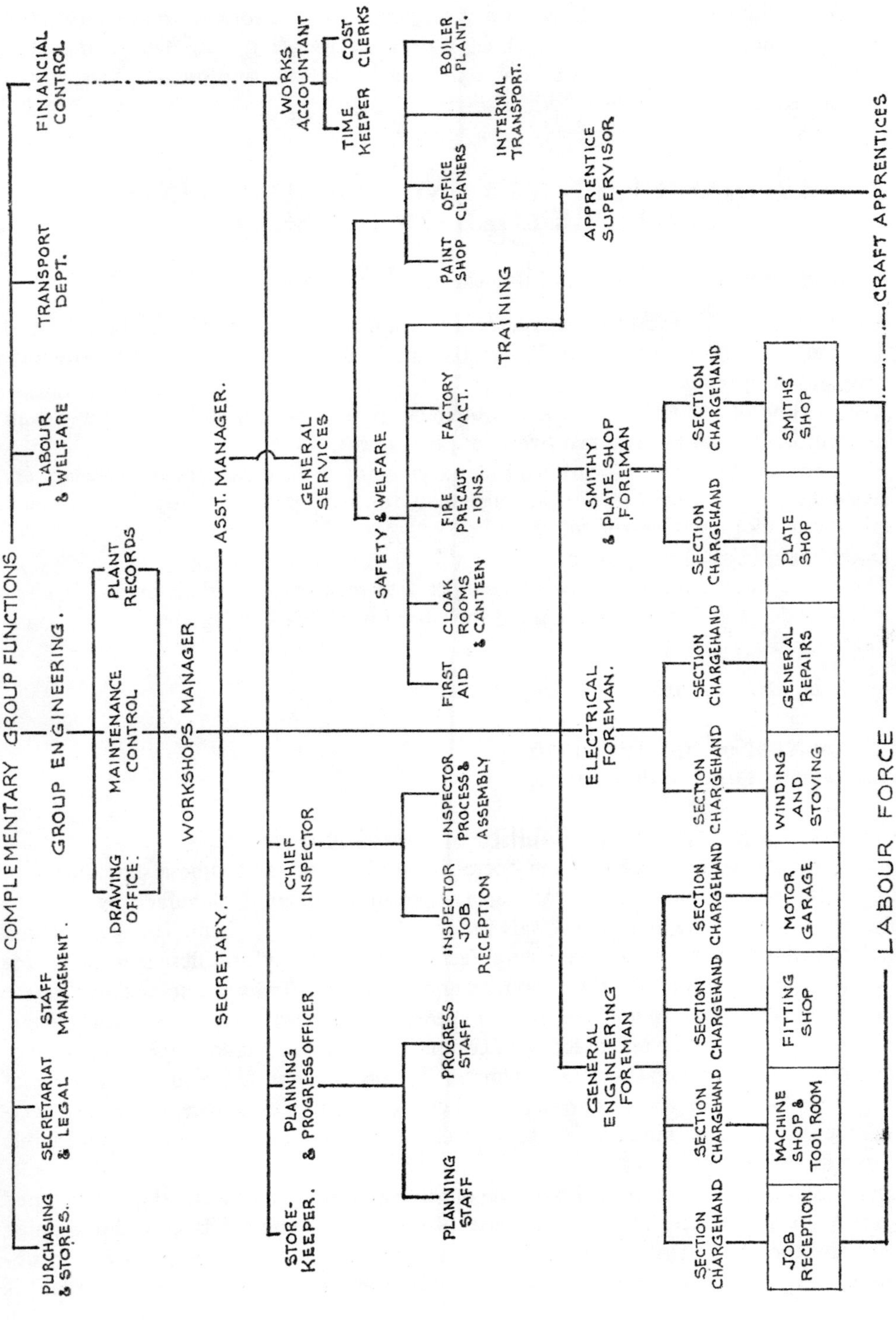


FIG. 4.9. Functional organisation and staffing chart—central workshops.

visual classifications as required. Supervisory personnel such as foremen and chargehands can be highlighted in this way. Personnel wastage, impending retirements, vacancies, prolonged incapacity, and so on can be visually signalled in the appropriate lines, so that the overall personnel position is readily assessed.

## BUILDING-UP, REGULATING AND MAINTAINING THE WORK LOAD ON THE SHOPS

The essential aspects of the control of the work-load are as follows:

The assessment of the nature, character, and volume of the work-load to be sustained.
The build-up and forward availability of spares and renewals to meet the developing programme envisaged.
The creation of a 'float' of plant and equipment to aid the required turn-round of plant in accordance with the planned overhaul programme.
Means of identifying and locating all plant, knowing the pre-determined overhaul and scheduling times, together with arrangements for securing their release for overhaul and controlling the turn-round.
Operational control procedures.
The gradual curtailment of all work being carried on away from the workshops which it is intended should become a part of their total work-load, being absorbed as conditions permit.

Considering these in turn:

### The assessment of the work-load

This is described earlier in this chapter.

### The build-up and forward availability of spares and renewals

This is important to the success of the operation, and at the same time a difficult factor to cope with. The problem lies in the dependence upon external manufacturers for the supply of a variety of spares and renewals for equipment made by them. The organisation and practices bearing upon 'after-servicing' facilities vary widely in different firms. There are two sides to this question—the supplier's and the user's. It is understandable that the manufacturer's organisation and activities are geared to the primary purpose of producing new equipment and that the manufacture of component spares and renewals is often secondary to that purpose. Similarly, where demand is irregular, the fixing and sustaining of unit stocks and spares levels does present problems to the supplier. Hence, it is often necessary for the user to accommodate a range of stocks from numerous firms, on an undesirable scale, due to delivery problems.

Experience in practice confirms this as one of the major problems facing those concerned with planned maintenance. The circumstances of new plant being available within a much shorter time than the delivery period for spares for servicing existing plant is not uncommon. The purchase of new plant to obtain components for servicing existing plant is both uneconomic and directly opposed to one of the main aims of planned maintenance; namely the deferring of capital outlay on new plant and equipment. The robbing of production lines engaged on making new plant, to meet urgent demands for components to service existing plant, is not compatible with successful planned maintenance. It can only be regarded as a palliative to the main problem—the lack of 'after-servicing' and spares from the maker.

From the user's point of view, the following should be aimed at:

The avoidance of unnecessarily high inspection standards, resulting in the premature scrapping of components.

The highest economic recovery of components for re-conditioning and re-use.

The establishment of quantitative, phased forecasts of demands, based on past trends and the programmes envisaged.

The maintenance of reasonable stocks, based on the best delivery conditions obtainable, and centralised so that the minimum holding affords maximum coverage. This is aided by the curtailment of work outside the central workshops.

The regulation of the demands and holdings so that they become based on recorded usage over appropriate periods, instead of relying on estimated forecasts and likely demands.

Giving manufacturers adequate and reliable forecasts of demands, as the level of work handled increases.

From the predicted programme of workshop throughput the Group's engineering organisation can draw up lists of initial spares and renewal demands for each plant category. The purchasing and stores department can then provide for their procurement in consultation with the various manufacturers.

*See over for Typical List*

The following is a typical list for a single category of plant unit:

ESTIMATED SPARES DEMAND

Manufacturer ..........................................................................

Machine or Plant ......................................................................

Machine Portion ......................................................................

Basis of Assessment: 48 overhauls/annum
4 overhauls/month

| Brief component description | Maker's Part No. | User's Stores Coding | Quantity per set | User's % demand | Usage quantity | Forecast period frequency (months) | Remarks |
|---|---|---|---|---|---|---|---|
| 1 | 2 | 3 | 4 | 5 | 6 | 7 | 8 |
| | | | 2 | 100 | 8 | 1 | Say 24 every 3 months |
| | | | 2 | 50 | 4 | 1 | Say 12 every 3 months |
| | | | 1 | $12\frac{1}{2}$% | 1 | 2 | Say 3 every 6 months |
| | | | 1 | – | – | – | Hold 1 in stock Replace on issue |

*Explanatory notes*

COL. 4 is the quantity of the component comprising a set for one machine.

COL. 5 is a measure of the frequency and scale of renewal, based on experience as overhauls are carried out; for instance, 100% indicates that a full set is found, from experience, to require renewal at each overhaul, $12\frac{1}{2}$% indicates that one eighth of this renewal rate is the average.

COLS. 6 AND 7 together comprise the estimated usage rate and demand; for example, taking the first entry, 4 machines overhauled per month calls for 8 components each month, as the % demand is 100. Taking the third entry, 4 machines overhauled per month calls for only 0·5 of that particular component each month, or 1 every 2 months. The fourth typical entry relates to the type of component not normally regarded as a wearing part, such as a housing or gear casing, but is one for which it is advisable to hold one spare in stock. This type of component would normally be on extended delivery, and holding the item in stock may be considered a sensible action.

COL. 8 is derived from Cols. 6 and 7, and gives a suggested practical basis on which the purchasing and stores department can negotiate with manufacturers and organise their coverage on the bases of agreed deliveries, with established maximum and minimum stock holdings ensuring availability to meet the overhaul programme.

Given these arrangements, continuous pressure is still often necessary to ensure that promised deliveries are met.

This approach can be applied to the spares and renewals programme for all plant, whether mobile, transportable or fixed, included in periodic scheduled overhauls and routine servicing.

Chapter 6 describes the relevant aspects of the stores and supplies organisation, and has a bearing on the matters here considered.

## *The plant float*

The purpose and assessment of the plant 'float' relates especially to the periodic overhaul of mobile or transportable plant units. It is essentially plant in circulation, and is a pre-requisite for the necessary turn-round of equipment in an established periodic overhaul operation. As plant in circulation, it is essential that once the pool is established, it must be maintained. Each item withdrawn must be replaced by its equivalent in exchange. If equipment is diverted from the pool to augment production as a result of an emergency, a replacement for the pool must be ordered immediately to restore the holding. It is also essential to establish the principles (laid down in the following sub-section) concerning the release and movement of plant. Any deviation reacts on the workshop performance and the turn-round that can be achieved, and has a cumulative detrimental effect on the standard of plant upkeep and reliability. The pool must be maintained if the necessary turn-round of plant is to be kept flowing.

## *Location and identifying of plant: pre-determined times and scheduling of overhauls and control of movement*

These are prime functions of the plant record and inventory section, which is considered in detail in Chapter 2.

Reference to the appropriate sections of Chapter 2 will help to clarify its functions in the present context. The records provide the identity and location of all plant. The history card provides for forecasting overhauls on the broad basis of monthly commitments over a period of three years. This is co-ordinated with the assessment of the work-load on the Group's workshops, which was developed in conjunction with it. The signals at the bottom of the plant history card detail the actual plant units, their location and their overhaul phasing (month and year). The record card provides the technical details which enable suitable replacements to be allocated in advance from the plant 'float', to secure the release of the plant in operation, for scheduled overhaul.

## *The operational procedure of control*

The operational procedure of control affecting the build-up, regulation and maintenance of the work-load on the workshops calls for co-operative action. The two essential features are first that all external work or service demands on the central workshops should be backed by works orders routed through and authorised by the Group's engineering office, and secondly that the latter should function as:

A central clearing house, clarifying data and instructions, providing required drawings and costing allocations, establishing priorities for effective completion; all meeting requirements of the Group as a whole.

A 'buffer', capable of absorbing and smoothing out the disturbances and irregularities arising from enquiries and demands made by the various Units, each of which may see only its own requirements in terms of service and priority. This serves to establish the genuine priorities from a Group point of view, and prevents the flooding of the workshops by enquiries and demands; it also prevents the introduction of a multitude of so-called priorities which only hinder a steady flow of work.

An authority spreading and phasing the work load of the Group, to achieve efficient and economic employment of the available services.

Monthly, the plant record and inventory section notify the maintenance control section of planned plant overhauls and scheduled servicing forming the work-load, in terms of specific equipment, location and plant numbers, to be undertaken during the following month.

Having noted at the beginning of the month the plant units due to pass through the central workshops for periodic overhaul the following month, the plant records office at the same time notes the identity, readiness and availability of replacement plant units from the 'float'.

Notification is then given by the plant records office (on the form shown in Fig. 2.6) to the workshop management, that the specific plant items are to be dispatched to the appropriate sites by the stipulated dates so as to release plant in operation, for receipt at the central workshops by the given dates. A copy of this notification is sent to the operating Unit or Units concerned.

The workshops dispatch the replacement plant items from the 'float' at the times stipulated, at the same time issuing plant movement and receipt advice forms (Fig. 2.4) to the plant record office, for consignments to each Unit. The monthly forward notification form is returned from the workshops to plant records when action thereon is completed. This serves as a double check at the records office on effected plant movements.

On receipt of the replacement plant the Unit sends to the plant record office a confirmation of receipt on a plant movement and receipt advice (Fig. 2.4), providing a further check on movement and location.

The operating Unit, working to the date already notified for receipt of the released plant item from the central workshops, effects the change-over and the dispatch of the released plant. When dispatching the operating Unit:

Labels the plant item (*see* Fig. 4.13).

Sends a works order for the plant overhaul to the Group's engineering office (*see* Fig. 4.12).

Sends a plant movement and receipt advice to the plant records office (*see* Fig. 2.4).

The released plant units are received at the Group's workshops job reception, and the 'float' is restored to the position necessary to ensure the smooth flow of overhaul work.

In order to operate a periodic overhaul programme successfully, the general procedure and stipulated time limts must be clearly established and adhered to. Where irregularities

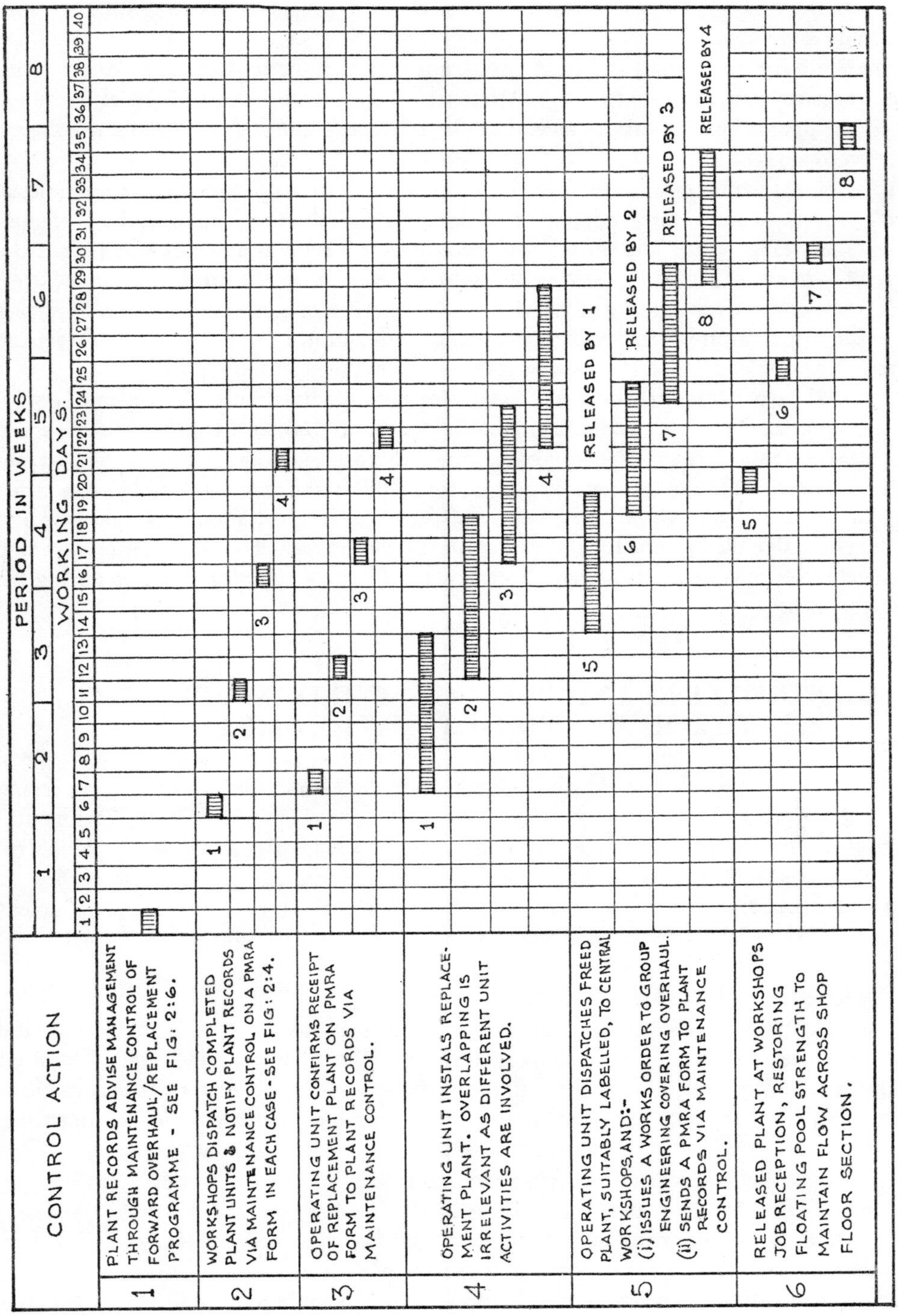


FIG. 4.10. Operational control and procedure chart for cyclic turn-round of one plant category.

occur, it is the function of maintenance control at the Group's engineering office to take appropriate action.

To maintain the turn-round and throughput, it is vital that the replacement plant is issued by the workshops on time, and that the stipulated period from receipt of the replacement plant at the operating Unit to the receipt of the released plant at the Group's workshops is not exceeded.

In this context is it helpful to refer back to Fig. 4.5 and to detail thereon the operational control, procedure and timing required to achieve the periodic turn-round in the particular plant category (Fig. 4.10). The shop floor operation for a section covering one plant category is illustrated in Fig. 4.11. Taking the second category in the work-load make-up—namely the routine inspection and servicing of incoming tackle and equipment—this is work arising from insurer's inspections to comply with statutory requirements. The repairs required to ensure continued insurance cover will be detailed in the reports.

The procedure is necessarily different to that for periodic overhauls. Each item of equipment can be treated as a plant item, and recorded on the Group and Unit plant record and history cards. At the Group's records office, the statutory inspections can be phased out on a basis acceptable to the insurer's inspector; for instance, the tackle at a certain Unit, in a particular geographical location, may all be inspected during the month of January, and these items are signalled as due for inspection during that month. The actual dates on which the inspections are to be made are confirmed by the Group's engineering office during the previous month.

## CONTROLLING AND PROGRESSING OF WORK INTO, THROUGH AND OUT OF THE SHOPS

It is useful first to outline the main points involved in the Group's organisation concerning the flow of a typical job from a single Unit. It is also necessary to consider the documentation required for the adminstrative and control functions involved. These should be regarded as typical and related to the particular organisational circumstances described. It should be remembered that paper work is only the means to an end, and every document must be devised to fulfil its specific purpose satisfactorily.

### The works order (Fig. 4.12)

This is the master authorising document issued by all sections of the Group, through the Group's engineering office to workshop management for all work and services required from the central workshops. It is the initiation of the workshop's action. Its instructions must be clear, precise, and complete. It should be possible for workshops management, on receipt of the order, to carry it through to satisfactory completion, without any need for back reference. If it is found necessary to request clarification the Group's engineering office have not fulfilled their function in this respect and remedial action should be taken. The pre-printed layout of the order is designed to aid clarity and conciseness and provides for the inclusion of all appropriate data in uniform style.

Users of works orders are issued with books of say 50 numbered duplicate orders (100

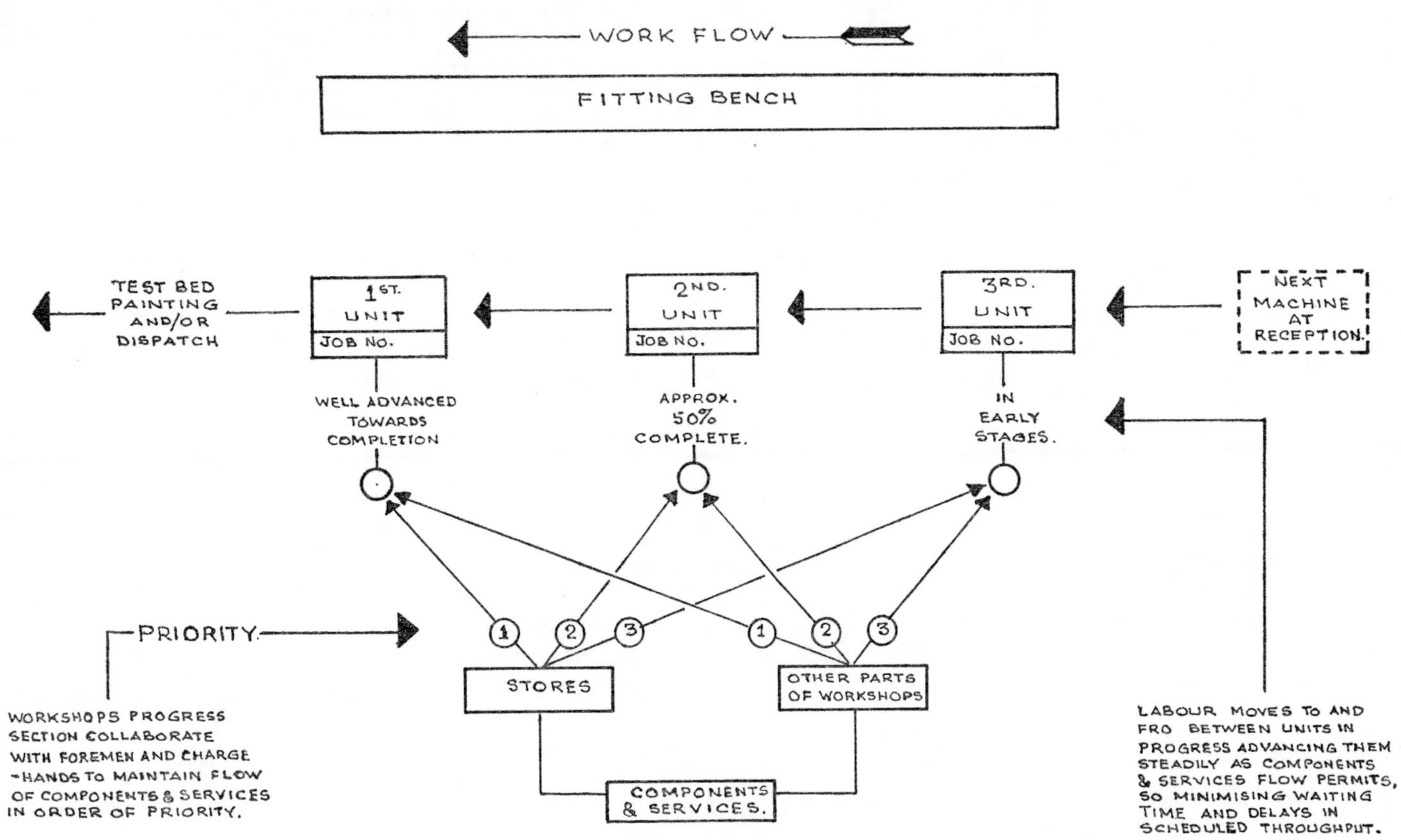


FIG. 4.11. Operational chart for shop floor section of one plant category—compressors.

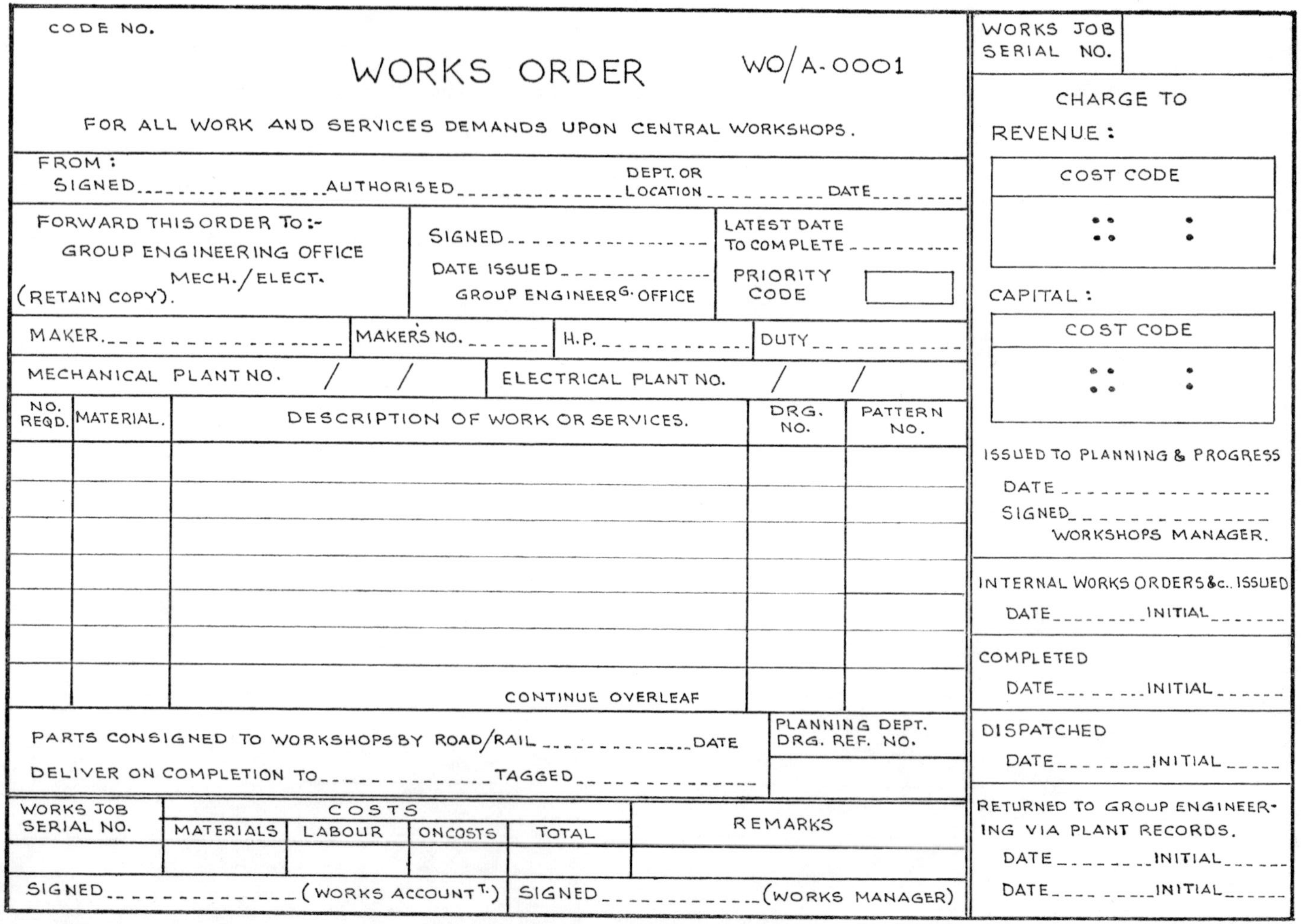

CODE NO.

WORKS ORDER WO/A-0001

FOR ALL WORK AND SERVICES DEMANDS UPON CENTRAL WORKSHOPS.

FROM:
SIGNED................AUTHORISED..............DEPT. OR LOCATION..........DATE........

FORWARD THIS ORDER TO:-
GROUP ENGINEERING OFFICE
MECH./ELECT.
(RETAIN COPY).

SIGNED................
DATE ISSUED............
GROUP ENGINEER G. OFFICE

LATEST DATE TO COMPLETE..........
PRIORITY CODE

MAKER................ MAKER'S NO. ....... H.P. ........... DUTY ...........

MECHANICAL PLANT NO. / / ELECTRICAL PLANT NO. / /

| NO. REQD. | MATERIAL. | DESCRIPTION OF WORK OR SERVICES. | DRG. NO. | PATTERN NO. |
|---|---|---|---|---|
| | | | | |
| | | | | |
| | | | | |
| | | | | |
| | | | | |
| | | | | |
| | | | | |
| | | CONTINUE OVERLEAF | | |

PARTS CONSIGNED TO WORKSHOPS BY ROAD/RAIL..............DATE

DELIVER ON COMPLETION TO..............TAGGED..............

PLANNING DEPT. DRG. REF. NO.

| WORKS JOB SERIAL NO. | COSTS | | | | REMARKS |
|---|---|---|---|---|---|
| | MATERIALS | LABOUR | ONCOSTS | TOTAL | |
| | | | | | |

SIGNED..............(WORKS ACCOUNT[T.]) SIGNED............(WORKS MANAGER)

WORKS JOB SERIAL NO.

CHARGE TO

REVENUE:

COST CODE

CAPITAL:

COST CODE

ISSUED TO PLANNING & PROGRESS
DATE..............
SIGNED..............
WORKSHOPS MANAGER.

INTERNAL WORKS ORDERS &c. ISSUED
DATE........INITIAL......

COMPLETED
DATE........INITIAL......

DISPATCHED
DATE........INITIAL......

RETURNED TO GROUP ENGINEERING VIA PLANT RECORDS.
DATE........INITIAL......
DATE........INITIAL......

FIG. 4.12. The works order.

sheets), with a tear-off top copy and the carbon copy being retained in the book (which can be indexed and filed for reference at the originating office). All works orders should be sent direct to the Group's engineering office, and not handed to lorry drivers or other persons to deliver.

Enquiries regarding the position and progress of any works order should normally be made through the Group's engineering office, quoting the works order number. Sketches should be on separate sheets, from the sketch books provided.

The latest date for completion is important, and should be genuinely assessed to aid the allocation of priority to incoming work. Requirements should be determined and advised as far ahead as practicable, so as to reduce urgent calls upon the workshops and assist the orderly working of departments.

The date and method of consignment to the workshops should be indicated when applicable; so should the consignment and labelling instructions on completion. All works orders should carry the initials of the person delegated as the issuing authority, or his representative.

In the case of urgent breakdown, the Unit concerned should contact the Group's engineering office, giving details to enable the work to be put in hand. The works order following should be marked URGENT BREAKDOWN—CONFIRMING TELEPHONE MESSAGE. The Group's engineering office issues to workshops management an interim note covering instructions and marked OFFICIAL WORKS ORDER FOLLOWING. It is important that this procedure is reserved for genuine breakdowns, and not used otherwise.

Where plant or materials are consigned to workshops against a works order, the latter must be sent to the Group's engineering office as early as possible, and in any event, not later than the day following dispatch of the plant or equipment. Here again, if slackness is allowed to develop, it has far-reaching consequences on operational efficiency.

On receipt of the works order at the Group's engineering office, it is examined together with any attached drawings or sketches, and checked for clarity of instructions, and to ensure that the work is properly authorised and can be approved. Any drawings or sketches required are cleared with the Group's drawing office and attached, certified, dated and signed by the issuing authority. Matters calling for investigation or discussion are cleared, reference back to the source of origin being made if necessary. The priority code reference is indicated, the charging instructions entered and signed by the Group's engineering authority, with date of issue to the workshops annotated.

When the order is accepted by workshops management, it is entered in a works order register and indexed under location or Unit, together with the works order number, date received, date issued, and brief description. The works order number provides a continuous job identification from the carbon copy at the source, through the Group's engineering office register, to the master works order at the workshops planning and progress office, and for dealing with subsequent enquiries as to the position of the job.

Where many jobs and types of work and services are involved, these procedures and practices will be found not only effective, but essential to achieve total control.

## Sketch books

These are issued to originators of works orders to prevent rough sketches being put on orders or odd scraps of paper (and to save time at Group level). The pages are numbered in duplicate with a tear-off top copy, and a carbon copy retained in the book which should be indexed and filed for reference. The detached sketch sheet should be pinned to the works order.

## The job label (Fig. 4.13)

This is attached securely to the plant item at the Unit; it serves to identify it on arrival at the Group's workshops and to correlate it with the works order (routed *via* the Group's engineering office) and the plant movement and receipt advice note sent to the Group's records office. It includes a brief description of the plant and its plant number, any accessories or fitments included being shown on the back. It also indicates the Unit or location sending in the plant and confirms:

The works order number and date of issue.
The plant movement and receipt advice number and date of issue.
The transport advice number and date of issue.
The signature of the consigning authority.

On receipt of the job at the Group's workshops, the detachable portion of the label is completed by entering the covering works order number and date, and the Unit or location from which the plant has been received. This portion is then sent to the planning office, confirming receipt of the job, and correlating it with the works order.

## The transport advice note (Fig. 7.1)

These are issued in book form to all points at which movements of men and materials are initiated. They are designed to give forewarning to the transport office of demands for transport. The pages are numbered in triplicate; the top two copies are perforated for extraction, the third copy being retained in the book, which should be indexed and kept for reference. The two top copies are sent to the transport office where details of the driver and vehicle are entered. The driver presents the two copies to the responsible person at the point of loading, who initials them to confirm the loading of the materials. Similarly, the responsible person at the point of unloading initials both copies, to confirm receipt; one copy is left at the point of delivery, and the other is returned to the transport office. This covers the transport office against any enquiries or complaints it may receive and also prevents dispute or confusion as to whether the plant was consigned, what the consignment included, where it was moved from and to, and where it is currently located.

## The plant movement and receipt advice (Fig. 2.4)

This is the document correlating all plant movements and receipts with the Group's plant records. They are issued in book form to all sources initiating plant movement or concerned with receipt of plant. The pages are numbered in duplicate, with a tear-off top copy and a carbon copy retained in the book, which should be indexed and filed for reference. Both the dispatching and receiving departments must complete the appropriate parts of a movement and receipt advice and forward it promptly to the Group's plant record and inventory office after consignment or receipt of the plant. This applies to plant units but NOT to

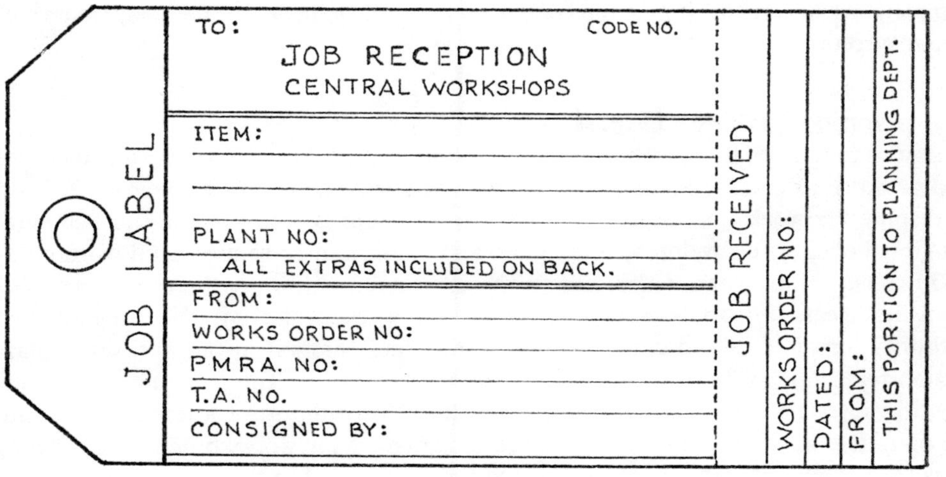


FIG. 4.13. The job label.

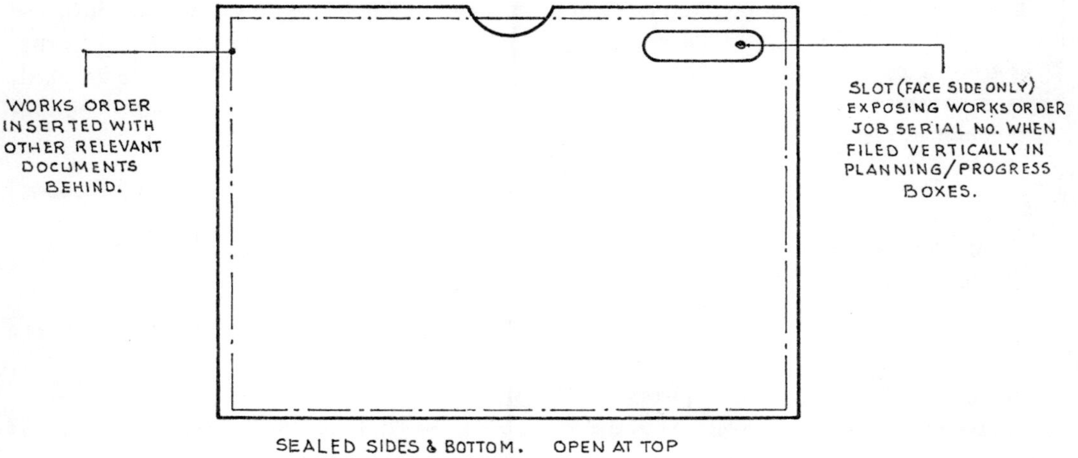


FIG. 4.14. The planning envelope.

component parts. The form should cater for every contingency arising from plant movement and receipt; by providing a clear layout for the entries and a minimum of hand-writing to complete the document, it aids the control of plant movement and the maintenance of the Group's records.

## The planning envelope (Fig. 4.14)

This accommodates the works order in the planning and progress office, together with all related documents involved in the planning, progressing, and costing processes. It is slotted to display the works job serial number when the order is filed vertically in planning and progress boxes. The envelopes, being plain and blank, can be re-used until worn out.

On receipt of the works order, and any accompanying drawings, the workshops management first clears any doubtful points with the Group's engineering office; the appropriate authority then initials and dates the works order and passes it to the workshops planning section for action, with any additional internal instructions. The planning section indexes any drawings and sketches included and files them for withdrawal when required, entering the planning drawing number in the space provided on the works order. This latter maintains the correlation between the order and the drawings. The works order is inserted in a planning envelope and placed in planning box 1 for attention, coloured clips being attached to the top edge to denote priority. This planning envelope and its contents follow the job progress through the planning and progress boxes as described later.

## The job card (Figs. 4.15, 4.16, 4.17)

This is, in effect, the identification label; it gives the job number, routing, operating instructions, and certification of inspection. It is prepared by the planning office and issued by the progress office and accompanies the job from receipt to completion. The back of the job card has a constant layout, but the front displays one of three forms and colours denoting particular movement or consignment conditions as follows:

A RED card is attached to jobs and components moving internally between sub-sections of the shops to a final assembly point.

A YELLOW card is attached to jobs or re-conditioned components intended for the Group's stores stock.

A BUFF card is attached to jobs or components to be routed directly to the Units.

## The internal works order (Figs. 4.18 and 4.19)

This is the planning document. They are made up in pads of, say, 50 numbered duplicates (100 sheets), gummed along the top edge for tearing-off. They are prepared in duplicate, the top copy being retained in the planning envelope along with the master works order (Fig. 4.12), in the planning office. The layout is designed to aid the recording of all essential planning and progress data and references likely to be involved or required in carrying through any type of job. The top copy is issued first to the costing section to indicate the start of a job with its job serial number; it indicates to the costs office where costs will arise, and acts as a check on the data coming into that office.

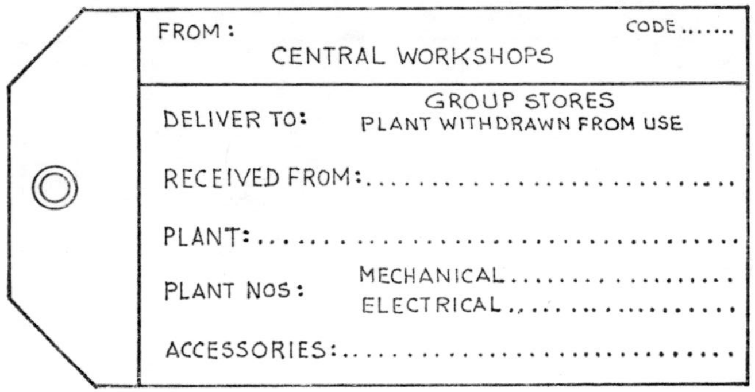


FRONT

FIG. 4.15. The job card (yellow)—for jobs and components from shops to stores. Rear layout as buff job card (Fig. 4.17).

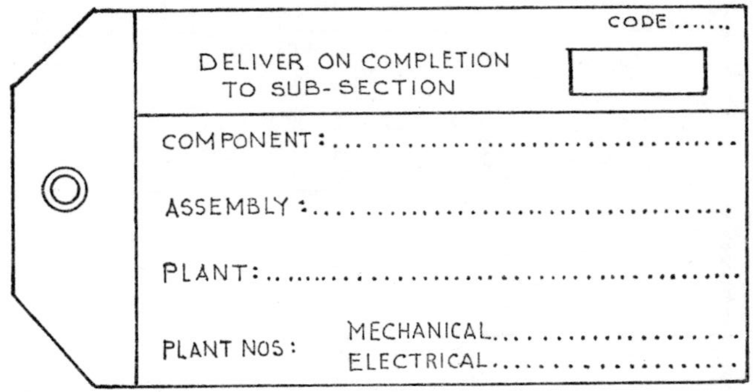


FRONT.

FIG. 4.16. The job card (pink)—for jobs and components circulating within the shops. Rear layout as buff job card (Fig. 4.17).

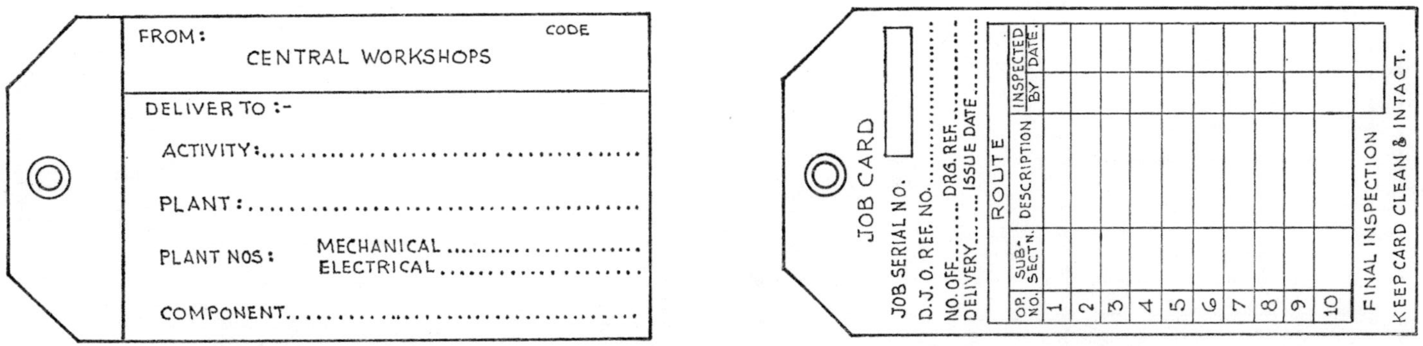


FRONT REAR

FIG. 4.17. The job card (buff)—for jobs and components from shops direct to unit activity.

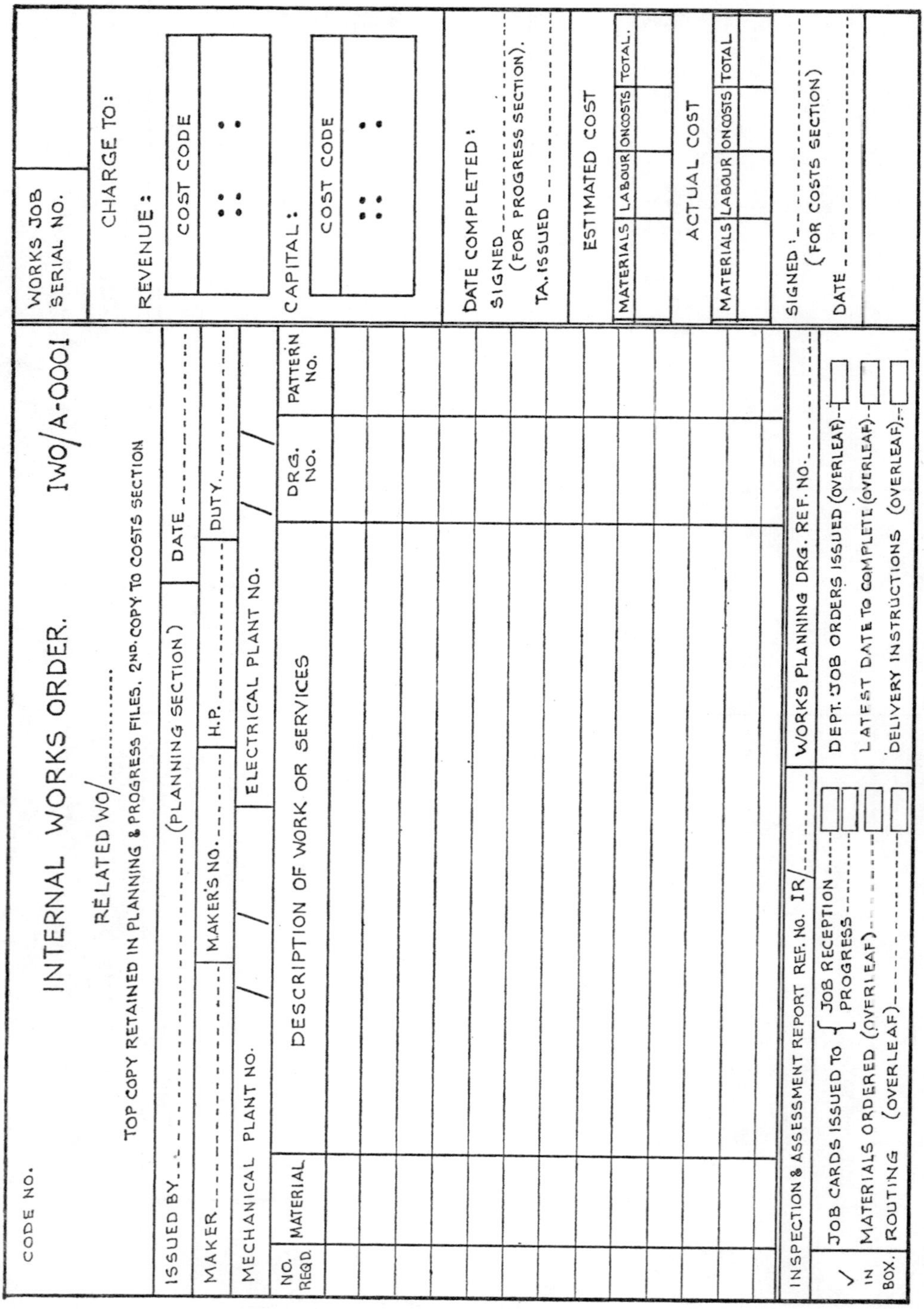

CODE NO.

INTERNAL WORKS ORDER. IWO/A-0001

RELATED WO/..........

TOP COPY RETAINED IN PLANNING & PROGRESS FILES. 2ND. COPY TO COSTS SECTION

ISSUED BY ---------- (PLANNING SECTION) DATE ----------

MAKER ---------- MAKER'S NO. ---------- H.P. ---------- DUTY ----------

MECHANICAL PLANT NO. / / ELECTRICAL PLANT NO. / /

| NO. REQD. | MATERIAL | DESCRIPTION OF WORK OR SERVICES | DRG. NO. | PATTERN NO. |
|---|---|---|---|---|
| | | | | |

INSPECTION & ASSESSMENT REPORT REF. NO. IR/---------- WORKS PLANNING DRG. REF. NO. ----------

✓ IN BOX.
JOB CARDS ISSUED TO { JOB RECEPTION / PROGRESS } ☐ ☐
MATERIALS ORDERED (OVERLEAF) ☐
ROUTING (OVERLEAF) ☐
DEPT. JOB ORDERS ISSUED (OVERLEAF) ☐
LATEST DATE TO COMPLETE (OVERLEAF) ☐
DELIVERY INSTRUCTIONS (OVERLEAF) ☐

WORKS JOB SERIAL NO.

CHARGE TO:

REVENUE: COST CODE

CAPITAL: COST CODE

DATE COMPLETED:
SIGNED ---------- (FOR PROGRESS SECTION).
T.A. ISSUED ----------

ESTIMATED COST

| MATERIALS | LABOUR | ONCOSTS | TOTAL. |
|---|---|---|---|
| | | | |

ACTUAL COST

| MATERIALS | LABOUR | ONCOSTS | TOTAL |
|---|---|---|---|
| | | | |

SIGNED: ---------- (FOR COSTS SECTION)
DATE ----------

FIG. 4.18. The internal works order (front).

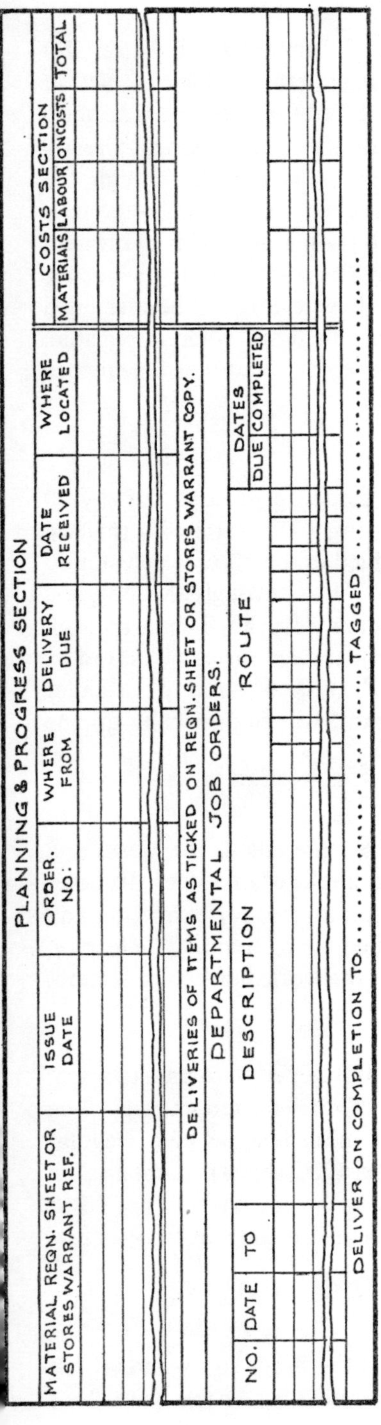


FIG. 4.19. The internal works order (rear).

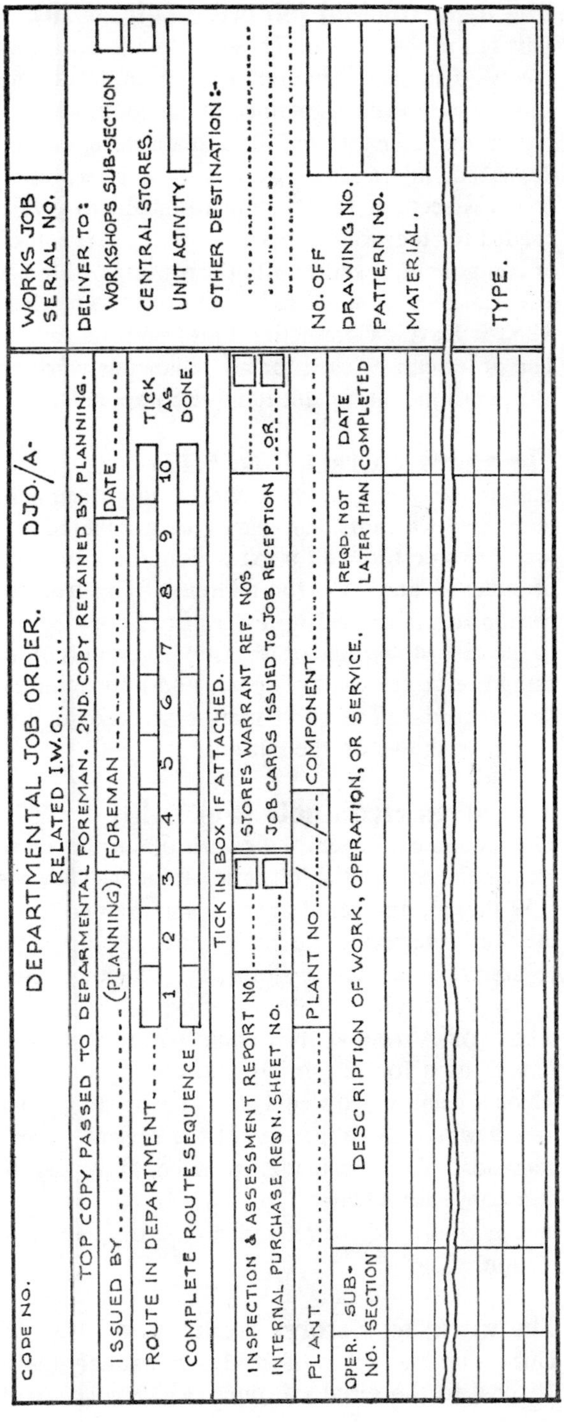


FIG. 4.20. The departmental job order.

## The departmental job order (Fig. 4.20)

This is the shop order. These are made up in books of, say, 50 numbered duplicates (100 sheets). The pages are numbered in duplicate, the top copy being perforated for extraction and the carbon copy retained in the book, which is indexed and filed for reference. The shop orders are made up from the data on the internal works order or planning document already described, all related planning data being shown. The top copy is issued, through the progress section, to the shop foreman, to give him all instructions, data, and references needed for the work in his section. It covers all of the operations on a particular plant item or component, together with the routing within his department. The complete route sequence also indicates from where he can expect to receive the item and to where it must be consigned when it leaves. Completion dates and delivery instructions are included when applicable. The order can be used for miscellaneous work and service demands, as well as a sequence of operations on a plant item or component.

## The stores warrant (Fig. 6.12)

This is the stores supplies and costing document. They are made up in books of say 50 numbered warrants. The pages are numbered in triplicate (giving 150 sheets) with the two top copies perforated for extraction, the third copy being retained in the book which should be indexed and filed for reference. The tear-off copies are issued through the progress section to the central stores, a priced copy being returned to the workshops costs section. In the case of periodic or repetitive overhauls of like units, considerable time can be saved by using pre-printed lists of spares and only entering figures for the quantity of any particular item required. The sheet can be secured to the stores warrant, which then calls up the materials shown on the parts list.

## The stores credit note (Fig. 6.13)

This document covers the return of surplus or wrongly-issued materials to the stores and credits the cost to the job serial number. They are made up in books of say 50 credit notes (150 sheets), numbered in triplicate, the two top copies being perforated for extraction and the third retained in the book, which should be indexed and filed for reference. The tear-off copies are issued to the stores, a priced copy being returned to the workshops costs section.

## The workshops stores issue note (Fig. 4.21)

This is used for the supply and costing of all consumables and general stores, stocks of which are consigned weekly to the shop store from the central stores. This reduces travelling time and avoids congestion at the central stores arising from miscellaneous and irregular demands. They are made up in books of say 100 issue notes (200 sheets), numbered in duplicate with a tear-off top copy and a carbon copy retained in the book which should be indexed and filed for reference. The top copy is priced and passed to the workshops costs section.

## The workshops stores credit note (Fig. 4.22)

This covers the return of surplus or wrongly-issued materials to the shops store, and credits the cost to the serial job number. They are made up in the same way as the workshops

| CODE NO. | | WORKSHOPS STORES ISSUE NOTE | | | | WSI/A 0001 |
|---|---|---|---|---|---|---|
| JOB NO: | | | DATE: | | | |
| CODE NO. | BIN NO. | DESCRIPTION | QUANTITY OR WEIGHT | FOR OFFICE USE | | |
| | | | | PRICE | PER | AMOUNT. |
| | | | | | | |
| | | | | | | |
| | | | | | | |
| | | | | | | |
| AUTHORISED BY | | | ISSUED BY. | | | |

Fig. 4.21. The workshops stores issue note.

| CODE NO. | | WORKSHOPS STORES CREDIT NOTE | | | | WSC/A 0001 |
|---|---|---|---|---|---|---|
| JOB NO. | | | DATE | | | |
| CODE NO. | BIN NO. | DESCRIPTION | QUANTITY OR WEIGHT | FOR OFFICE USE | | |
| | | | | PRICE | PER | AMOUNT. |
| | | | | | | |
| | | | | | | |
| | | | | | | |
| AUTHORISED BY | | | RECEIVED BY | | | |

Fig. 4.22. The workshops stores credit note.

| CODE NO. | OPERATION SLIP | | |
|---|---|---|---|
| D.J.O. REF. NO. | | WORKS JOB SERIAL NO. | |
| PLANT | PLANT NO. / / | COMPONENT. | |
| NO. OFF: | DRG. REF. NO: | CHECK NO. MACHINE NO. | DATE |

| OPER. NO. | SUB-SECTION | OPERATION : | TO SUB-SECTION |
|---|---|---|---|
| | | | |

| RECEIVED | PASSED | REJECT. | TO RECTIFY | INSPECTION | DATE | PROGRESS | TRANSFERRED |
|---|---|---|---|---|---|---|---|
| | | | | | | | |

Fig. 4.23. The operation slip.

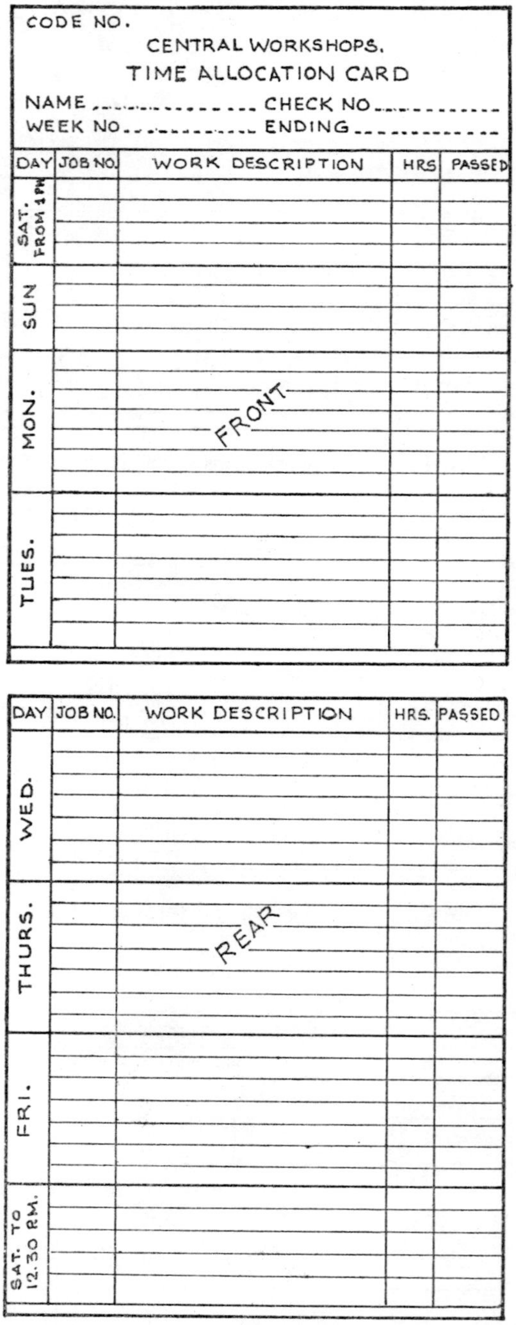


FIG. 4.24. The time allocation card.

stores issue note and are of similar arrangement. The top copy is priced and passed to the workshops costs section.

### The operation slip (Fig. 4.23)

This is the instruction to the operative on the shop floor. They are made up in tear-off pads of say 100 sheets, and give instructions, direction, and action requirements between the foremen, chargehands, and operatives to suit the progressing and time allocation. They also define the order in which the operative deals with the jobs allocated to him.

### The time allocation card (Fig. 4.24)

This is the document used to apportion the shop labour time against the various job numbers. The hours entered are checked against the operative's clock card (see next sub-heading). They are issued weekly by the time office; detailed directions on its use are given later.

### The clock card (Fig. 4.25)

This matches the time-recording equipment at the time office. It records the total hours worked, daily and weekly, by each individual, against his check number. This card must be reconciled weekly with the time allocation card.

### The completed job record envelope (Fig. 4.26)

This is used in the progress section to assemble all the documents relating to each completed job. The workshops management and other appropriate authorities within the engineering organisation can call these in for scrutiny; those considered useful for future reference may be segregated and suitably filed.

## SUMMARY OF DOCUMENTATION WITH STATIONERY CODINGS AND PRINTING INSTRUCTIONS

In order to ease the introduction and maintenance of stocks of the foregoing documents and stationery, a tabulated summary should be prepared. This indicates to the stationery department the initial quantity for ordering, printing instructions, and suggested maximum and minimum stocks. The stationery department enters the stationery code reference on the summary, and adds it to all documents employed by the user department when further supplies are required. It is particularly advisable in the early stages of a scheme to work on a short forward basis of say three to six months; if modifications are then found to be desirable, only small stocks of stationery are scrapped.

## OPERATIONAL PROCEDURE AND CONTROL

A composite functional and operational procedure chart should be prepared, to include initiating sources and authorities, and illustrating the flow of documentation, together with an indication of action, where this is necessary to clarify procedure. This gives a visual

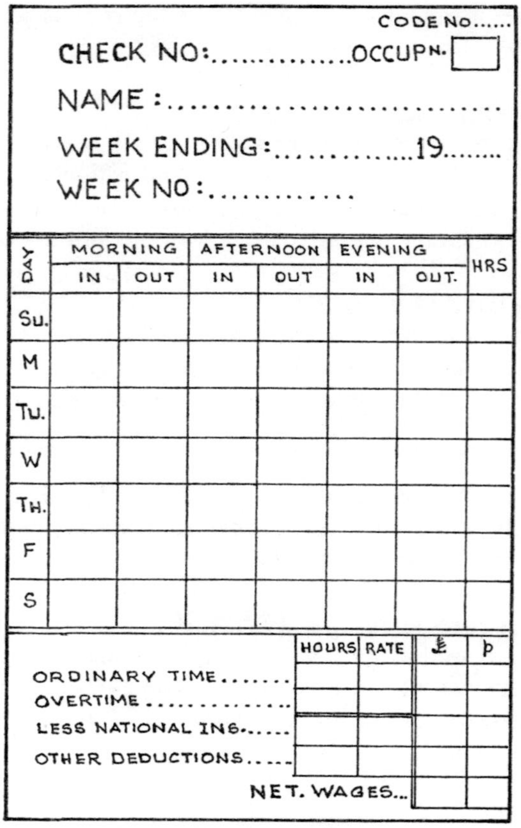


FIG. 4.25. The clock card.

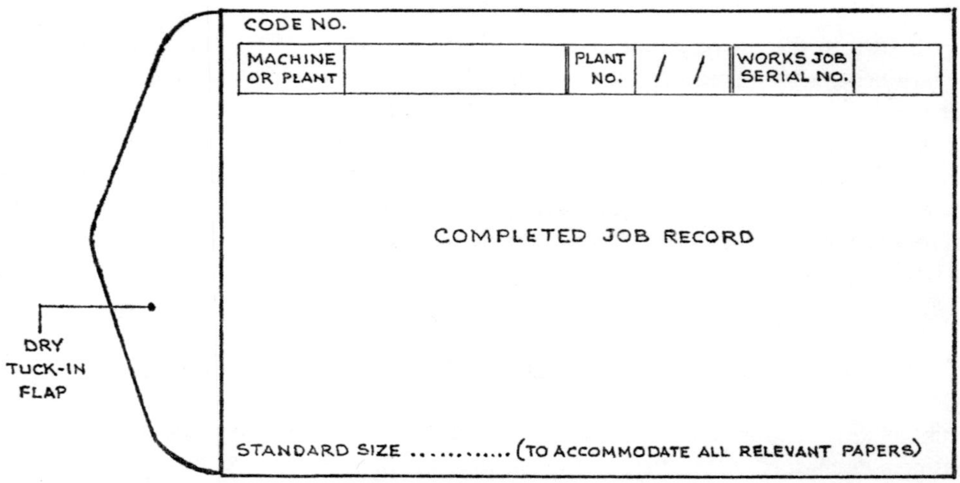


FIG. 4.26. The completed job record envelope.

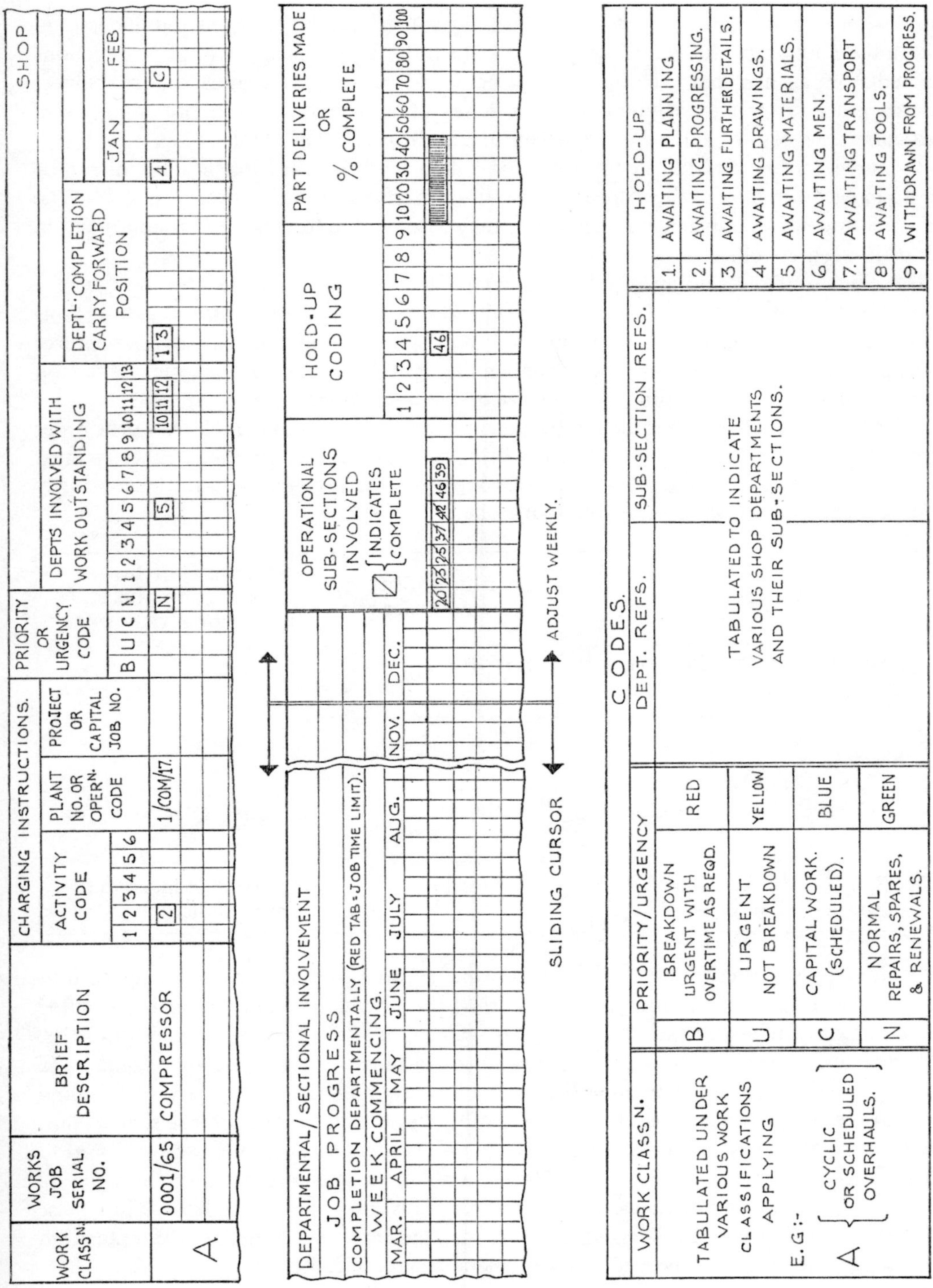


FIG. 4.27. Job progress board—central workshops.

impression from which directions for all initiating sources can be prepared and understood within the context of the whole. Appendix I illustrates a specimen set of such directions derived from such a chart. To aid management and control, a composite Job Progress Board, using proprietary equipment, or a series of such boards each providing for a particular category of work, can be set up in the Progress Office (Fig. 4.27).

To augment this, and to aid control on the shop floor, forward loading boards can be set up in each foreman's office, covering the sub-sections under his control (Fig. 4.28). The progress function is implemented departmentally through the foreman and is controlled in the main progress office.

The control of the routing of jobs and components is greatly assisted by the use of a block layout diagram (Figs. 4.8 and 4.8(a)) identifying the shop sections and sub-sections with their operational functions. Copies of the diagram should be displayed wherever appropriate. This helps to channel the work in the right directions, to keep track of the job location, and to speed and simplify the work of the internal transport.

A further essential feature is the establishment of priorities. From the master works order (Fig. 4.12) and the explanatory notes, it will be seen that the originating authority must enter the LATEST date for completion. The Group's engineering office add to this a priority code reference; a typical priority coding may be as follows:

| Category | Enter in Priority Coding Space | Enter in Date for Completion Space (if not already shown) |
|---|---|---|
| URGENT BREAKDOWN with overtime as required | **B** (Red) | Immediate |
| URGENT ATTENTION—not breakdown | **U** (Yellow) | Date required by |
| CAPITAL WORK—scheduled | **C** (Blue) | Estimated latest date for completion |
| NORMAL REPAIRS, SPARES, AND RENEWALS | **N** (Green) | Approximate latest date for completion |

The Group's engineering office should ensure that all works orders transmitted or issued by them, are given one of the foregoing categories; the originator of the order is consulted where necessary, to enable the workshops management to retain a clear picture of priorities. It is vital that the priorities awarded are genuinely attuned to the true requirements, and that urgent calls and demands upon the workshops are minimised. Disregard of this, and abuse of the coding, only lowers the standard of the service given. In a scheme such as this, it is essential to clarify to all concerned both the parts they play within it, and the need to conform to the procedure laid down. Hence, a set of directives and instructions for all initiating sources should be issued; a typical directive is shown in Appendix I. While these must be set out in some detail, common sense must be employed to avoid inflexibility and the stifling of the initiative needed to overcome unusual difficulties.

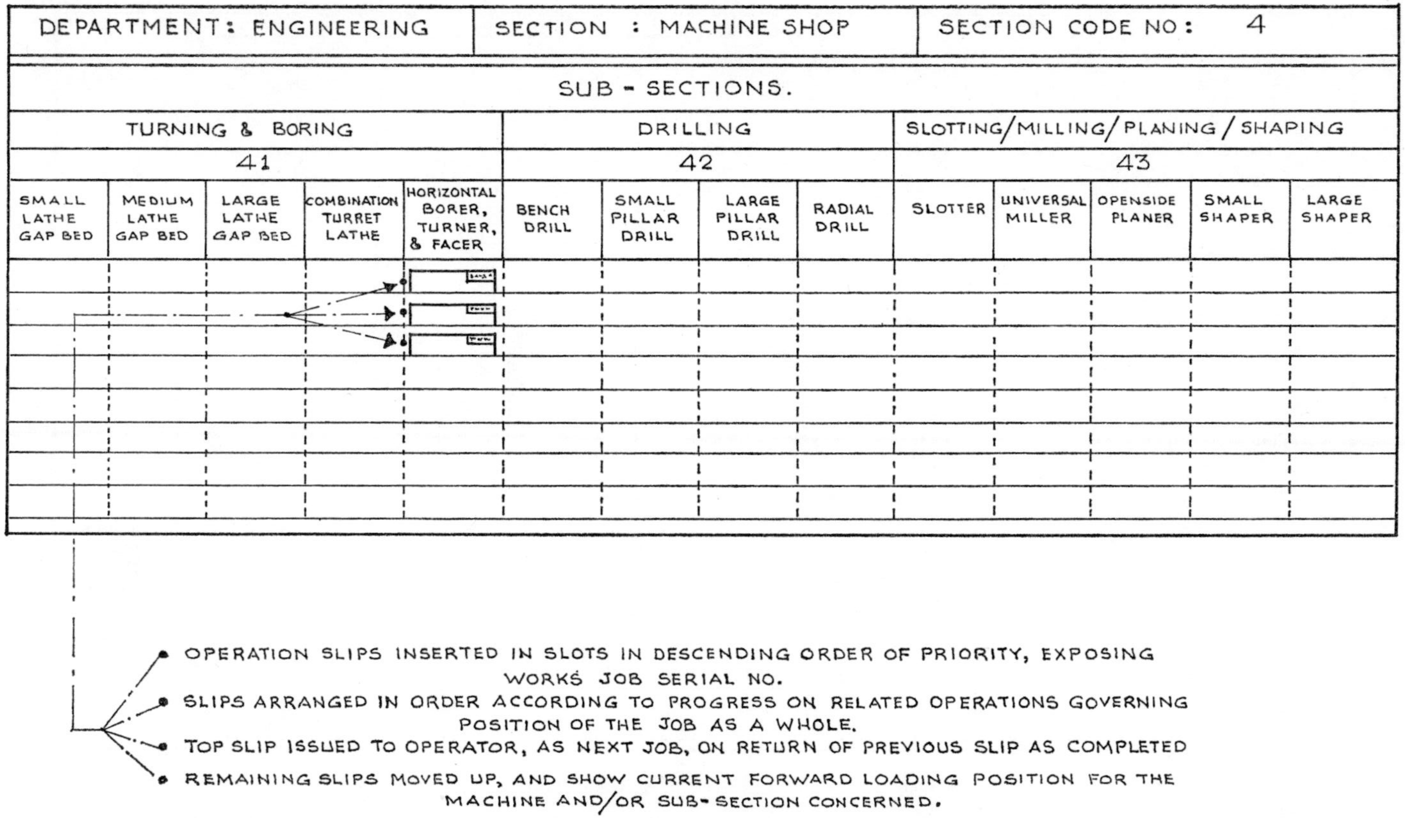


FIG. 4.28. Sub-sections forward loading board—central workshops.

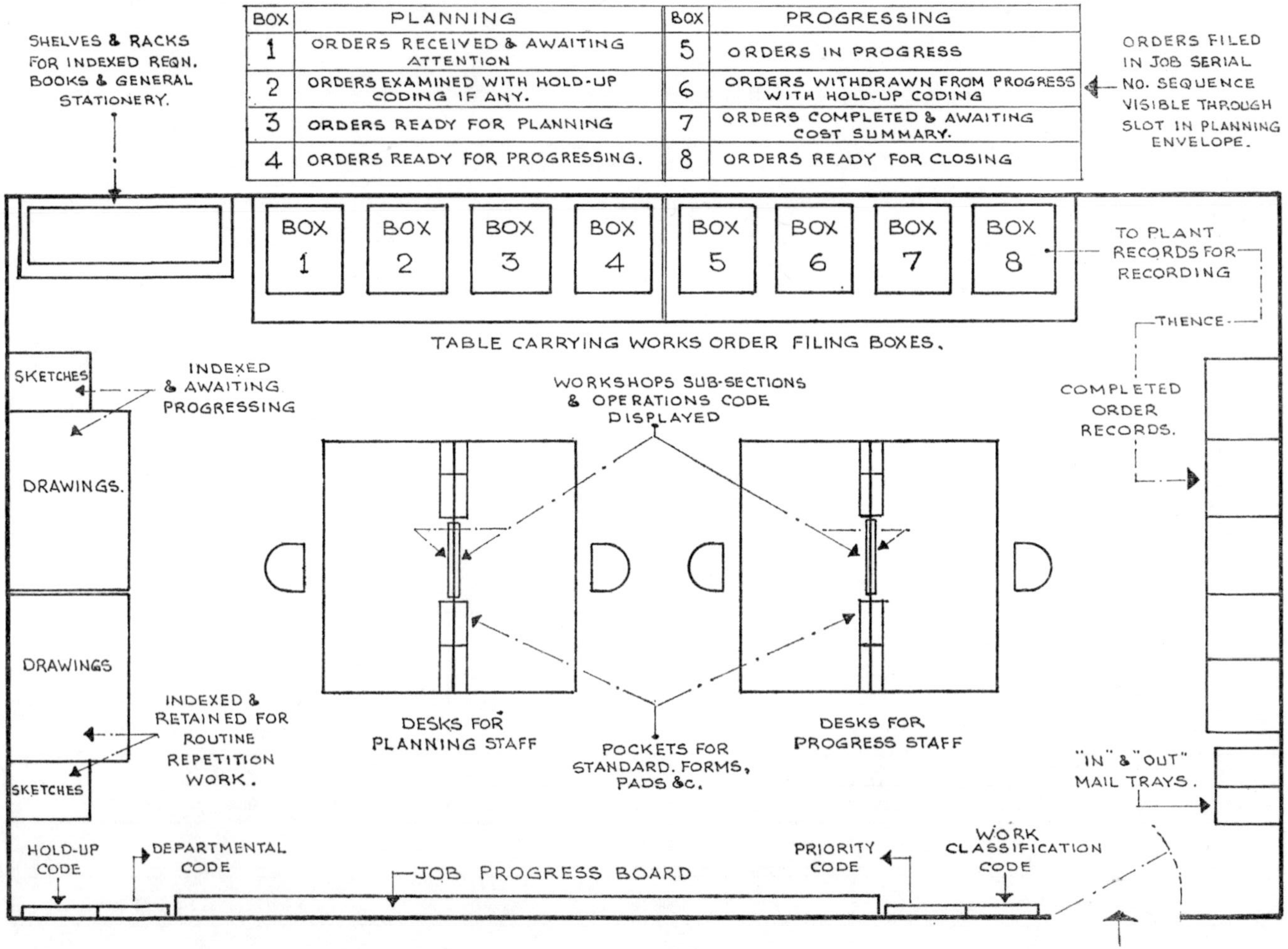


FIG. 4.29. General arrangement of planning and progress office—central workshops.

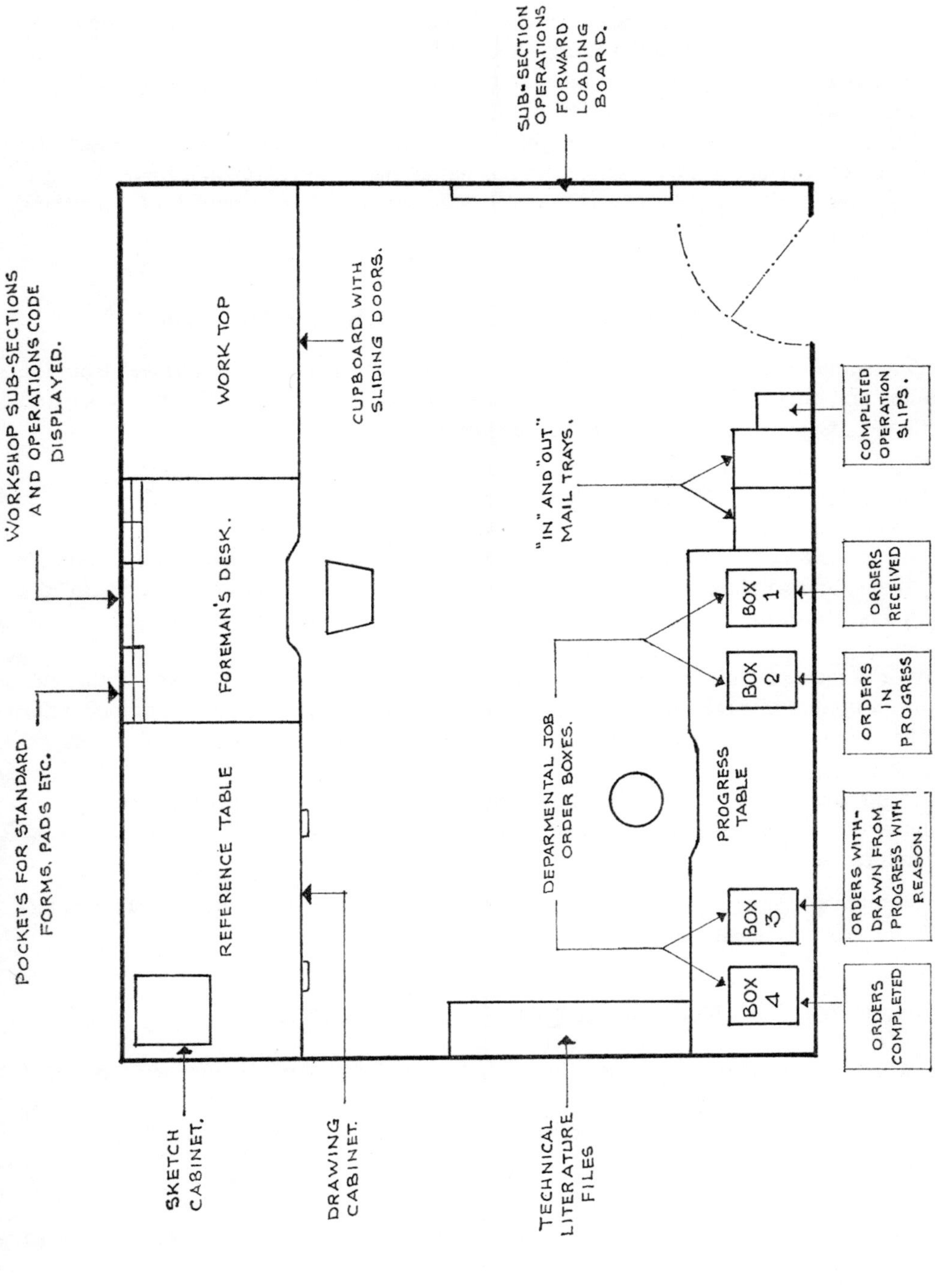


FIG. 4.30. Workshop foreman's office lay-out.

## THE CONTROL ACCOMMODATION

The accommodation and arrangement of the planning and progress office, the time office, the cost office, and the foremen's offices should be considered in relation to staffing, the duties and operations involved, and the system of control employed. Operational convenience, and the elimination of unnecessary effort and movement of personnel are also important. This can be helped by providing suitable internal communications between the sections; modern inter-communication systems meet this requirement, and can make a useful contribution to overall efficiency. Figures 4.29 and 4.30 show typical arrangement of a planning and progress office and a foreman's office.

## CLEANLINESS AND ORDERLINESS

Cleanliness and orderliness are important aids to efficiency. This must be stressed from the beginning and once the means and aids have been provided, high standards must be set and maintained. In this connection, the following are relevant:

The provision and regular use of apparatus for white-lining the shop floors.

The demarcation of the main walkways (and keeping them clear).

The clear demarcation of shop sections, with their identity and purpose.

The provision of adequate internal transport and auxiliary equipment to keep materials moving, and to place them properly in position. Overhead cranes, battery trucks and trailers, slings and cradles, pallets and stillages, and containers are all part of this service.

Arrangements for refuse and scrap collection and disposal should be made. Litter bins and containers for the various types of scrap metal should be located at suitable points.

The cleaning and lubrication of all shop equipment should be a matter of established routine.

Arrangements should be made for the regular cleaning and tidying of offices.

The use of a pleasant colour scheme throughout the shops and offices, and its regular maintenance.

The directives and instructions to the initiating sources (referred to earlier in this chapter) should include a reference to this heading.

## FURTHER CONSIDERATIONS

Once these arrangements are in operation, it is possible to reassess particular aspects and consider proposals for improvement, in the knowledge of the repercussions which these would have outside their fields of application. Such considerations in particular sections might include:

Time and motion studies, to minimise the labour effort and maximise its output.

Method studies, including the standardisation of methods, the introduction of special tools, jigs, and fixtures, the improvement of processes, and so on.

Improvements in the standards of workmanship and inspection.

The introduction of methods and processes to reduce clerical work.

The establishment of standard times and costs for specific tasks, determined from experience, to allow comparisons of performance and to indicate those cases where deviation appears to merit investigation.

In all of these considerations, the need for increased efficiency should not be pursued to the point at which initiative is repressed and human relationships suffer, to the detriment of the work.

*Chapter 5*

# INSPECTION AND MAINTENANCE OF SITE PLANT

## Organisation and Control

## INTRODUCTION

There is a distinction between the conditions applying at the Unit and those at the Group's workshops. The latter are essentially a service activity, whereas the Unit is a production activity.

At the Group's workshops, many types of machinery and equipment, in varying condition, are being handled, and a wide range of services is called for. These workshops must not be regarded as an entity; the true measure of their efficiency is the degree of success with which they sustain production and meet its demands throughout the Group, while preserving the material assets of the Group. While some of the work comes into the category of periodic overhauls, entailing forward planning, it should not be regarded as truly repetitive work, or essentially to be carried out to a rigid, long term plan. Plant condition or overhaul needs depend markedly on operating conditions, usage, and treatment, while age bears upon design and interchangeability. This type of work should rather be dovetailed into the overall activities in such a way that the desirable rates of throughput are attained on average over the year, without undesirable repercussions in other directions (*see* Chapter 4). Close and continuous direction and supervision in the various sections of the shops, with scope for the exercise of judgement and initiative by the foremen and chargehands, and for their collaboration, should be encouraged; this ensures that the shops are attuned to the needs of the Group, and not conversely. At the Units, the work is mainly concerned with the inspection and maintenance of static plant and equipment directly sustaining production. A plant item released for periodic overhaul is replaced by its equivalent. Re-sited plant remains within the control of the Unit. Redundant or obsolete plant is removed from the Unit for transfer or scrap. Additional plant is brought under the control of the Unit. In these senses, the Unit can be regarded as a producing entity, with its own servicing resources which are not affected by occurrences outside the Unit, and are supported by the Group's resources. These conditions are much more favourable to forward-scheduling of work, tasks, and labour deployment. The time factor, and adherence to prepared schedules for the separate jobs become more important, since they have a direct effect upon production. Direction and supervision cannot be so close as that at the Group's workshops, because of the disposition of the plant and the separate locations involved; this also supports the case for forward-scheduling. Further, the jobs and tasks are more repetitive than those at the workshops, and forward schedules covering specific work, facilities, and labour resources can largely be retained, once they are established, provided that they are adjusted to take account of experience, and to allow for plant deletions and additions. Separate provision can be made for irregularities and unforeseen occurrences, calling on the Group's resources if necessary, and leaving the scheduled work proceeding to plan.

## GENERAL ORGANISATION ON SITE

Figure 5.1 shows a typical functional chart for a Unit, with supporting Group functions and services. In Fig. 1.4, the Group's engineering services are depicted, but in the present case a Unit's engineering service is given in more detail, together with Group's services and functions related to it. The following features should be noted:

The Unit plant and equipment is basically either mobile or transportable, or fixed.
The Unit's engineering services embrace three basic aspects:

*Plant Maintenance:* cleaning, lubrication, painting, inspection, routine servicing, running repairs and adjustments including breakdowns, and site overhauls.
*Operational:* control of operation and operational staff, in such cases as power stations, boiler plants, locomotives, and so on.
*Installation:* periodic replacement of mobile or transportable plant by reconditioned units. Additions and modifications to plant and equipment.

All of these services can be pre-planned, with the exception of running repairs and adjustments outside regular servicing routine, arising from inspection; breakdowns; and additions or modifications, which will call for special scheduling and programming, being irregular and varied.

To help meet these needs, the Unit may be supported in the following ways:

It has its own small workshops with limited facilities for 'first-aid' repairs and servicing; these minimise production losses and relieve the Group's workshops of minor work.
It has its own plant record and inventory, and maintenance control centres, accommodated near to the engineer's office and the workshops.
It is supported by the Group's engineering organisation in the heavier and more specialised tasks, in unforeseen emergencies, and in work related to the Group as a whole:

*Through the Group's workshops:*
Shop work for site overhauls.
Labour services on site overhauls and breakdowns.
Shop work on miscellaneous plant and components, repairs and replacements, servicing of equipment.
Provision of mobile or transportable plant units to replace those at site for periodic overhaul.

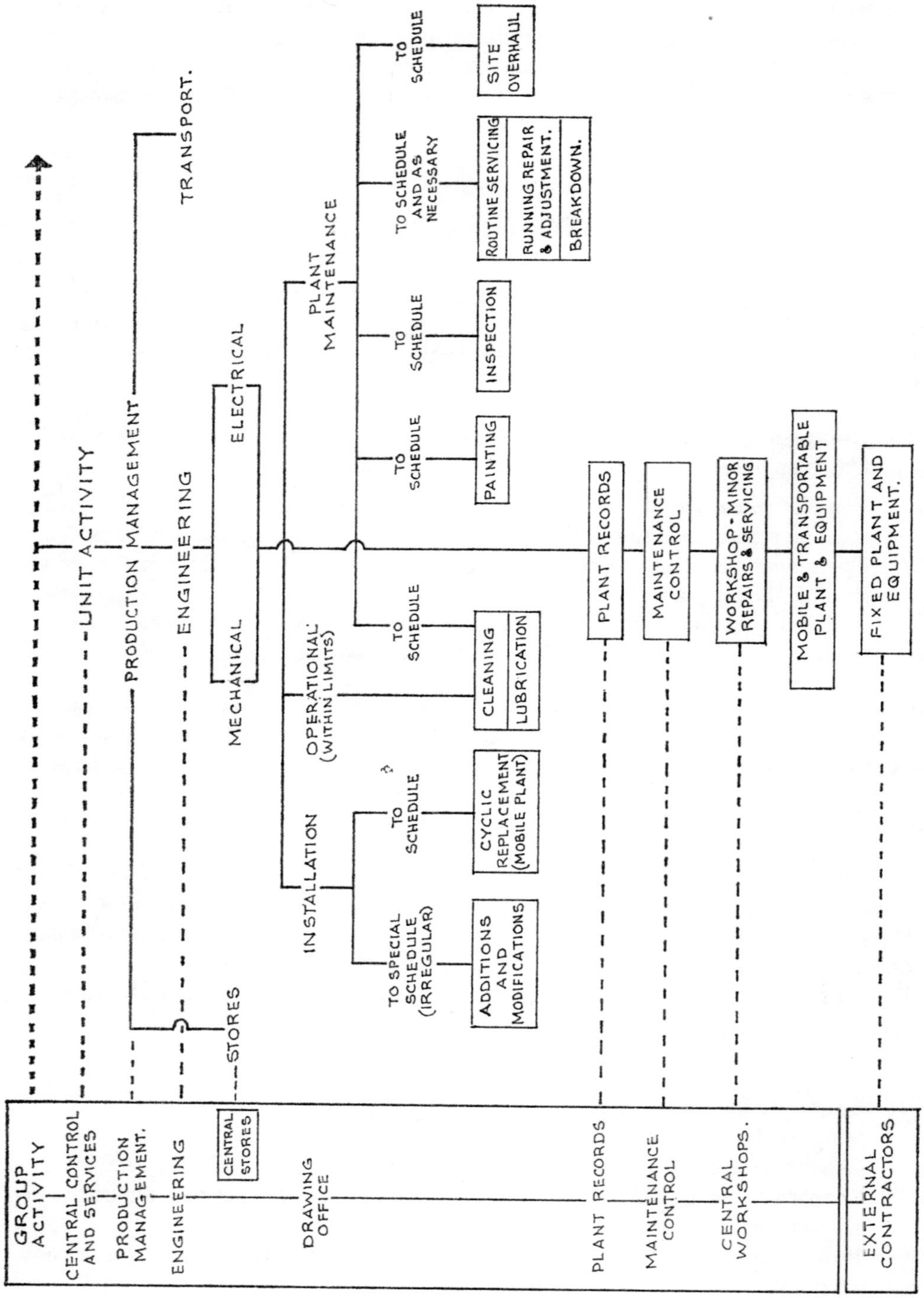


FIG. 5.1. Typical unit activity functional chart highlighting engineering.

Supply, on hire, from pool of special tested and certified tackle and equipment (mobile crane, transportable loading gantry, winches, lifting blocks, slings, shackles, etc.).

Provision of additions or modifications to plant and equipment under reconstruction or development schemes—to be limited to work incidental to such schemes, and those portions not covered by external contracts.

Labour services for above, with some limitations.

*Through the Group's plant records and maintenance control:*

Correlation of the Unit's records with those at the Group, so that there is co-ordination of the functions outlined in the plant record and inventory section (Chap. 2).

Correlation of the Unit's maintenance control with that at the Group, in establishing schedules for the various categories of work, ensuring that the Unit's schedules are suitably phased (where they form part of composite Group commitments), and fulfilling the maintenance schedules as a whole.

Keeping records of maintenance and furnishing the data from which performance can be judged.

*Through the Group's engineering office (including the drawing office):*

In matters calling for technical investigation or specialist advice.

In the supply of drawings and sketches, specifications and performance data, standards data, and general technical and operational or maintenance data.

In the exchange of ideas, experience, and practices within the Group.

In the provision of testing facilities and personnel.

In the provision of civil engineering services when required.

In the engagement of external contractors and facilities to augment or replace internal facilities and labour on site overhauls, breakdowns, and repairs and servicing of plant and equipment.

Provision of additions and modifications to plant and equipment.

There is also the Unit's stores, which are correlated with the Group's stores organisation in respect of supplies of consumable stores, and also of the provision and storage of spares and renewals. These may be priority items, to be stocked at site for emergency; or required for replacements and overhauls carried out at site, as distinct from those carried out at the Group's workshops; with the proviso that major items should be stored at the Group.

It has the services of the Group's transport organisation.

In matters concerning staff and labour, costing and finance, secretariat and legal aspects, it can call upon the resources of these departments of the Group's organisation.

## FURTHER PRELIMINARY CONSIDERATIONS

In addition to the foregoing, some further preliminary considerations should be taken into account.

It will be evident that the variety of tasks to be performed makes the adoption of some form of priority scale essential. Such a scale may take the following form:

| | | |
|---|---|---|
| STATUTORY | Work arising from statutory requirements | Must be done |
| INTERNAL MANDATORY | Work arising from internal mandatory directives | Must be done |
| DESIRABLE | Deemed necessary to attain the overall standard required | To be aimed at |
| UNPREDICTABLE | Non-scheduled work in above categories | A diminishing feature |

The scheduling and control of some categories of site maintenance work should preferably be treated on a Group basis, rather than a Unit basis. These categories are as follows (*see* Fig. 5.1):

## For scheduling and control on a group basis

Work in connection with the fulfilment of statutory requirements which involves inspection and certification by the appropriate insurance company's surveyor or is related to visits by government inspectors and their subsequent reports. Apart from the need for Group control over matters which may involve legal aspects, the Group basis of control enables the phasing of inspections to be scheduled to suit the convenience of both the insurance company and the undertaking.

Major overhauls of fixed plant at site, so that these can be phased to spread the load on Group services and external resources, and also to optimise the use of Group services before employing external aid.

Cyclic replacements of plant units, so that this important section of the Group's workshops activity proceeds as smoothly as possible on the basis described in Chapter 4.

Work related to pool services of the Group, such as a locomotive fleet, a fleet of road vehicles, or pool tackle and equipment.

Work involving the employment of resources external to the Group. This is essentially under Group control to ensure the most economic use of internal resources before employing external aid.

Work related to reconstruction and development; strictly speaking this is not maintenance, but it may affect the resources available.

## For scheduling and control on a unit basis

Cleaning; lubrication; painting; inspection; routine servicing; as distinct from major overhauls.

## Unscheduled work controlled on a unit basis

Running repairs and adjustments arising out of inspection, and breakdowns, with Group assistance on request.

To mould all these various factors into an efficient site maintenance service, truly equal to its tasks, is no mean achievement.

## GENERAL DEVELOPMENT OF ORGANISATION, PROCEDURE AND CONTROL

Figure 5.1 illustrates a typical site maintenance organisation, and the supporting Group organisation. Consideration is next given to the general development of organisation, procedure, and control for site maintenance. As a first step, it is advisable to break down the task into its elements by noting the questions it brings to mind:

What equipment has to be dealt with?
Where is it?
What needs to be done to it?
When is it to be done?
What periods are required?
What periods are available?
What men are required?
What men are available?
How may they be deployed to the best advantage?
How are they to be directed and instructed?
How is their time to be allocated and charged?
How should the necessary stores, spares, and renewals be controlled, supplied, and charged?
How is transport of men and materials to be ensured?
How is the work done to be recorded, and outstanding work kept in mind?
What general site facilities and labour resources are required?
How are these facilities and resources to be reinforced when necessary?
How is overall planning, operation, and control effected?
What are the costs of maintenance?

If the procedure is to be effective and efficient, it must provide complete answers to these questions. While they are set down in logical sequence they cannot be resolved in the same order, as each involves factors which react on the others. They are therefore useful in setting down the main lines of approach to the task, rather than as matters to be resolved in turn. These lines of approach are as follows:

The establishment of the basic personnel and functional organisation, derived from a typical Unit chart (Fig. 5.1).

The assessment of the Unit's engineering work load in terms of the location, categories of plant, nature of work, jobs, and tasks, together with their period, frequency, and phasing, and their labour demands in the various trades.

Preliminary scheduling, taking account of Group control (where applicable) and of plant availability.

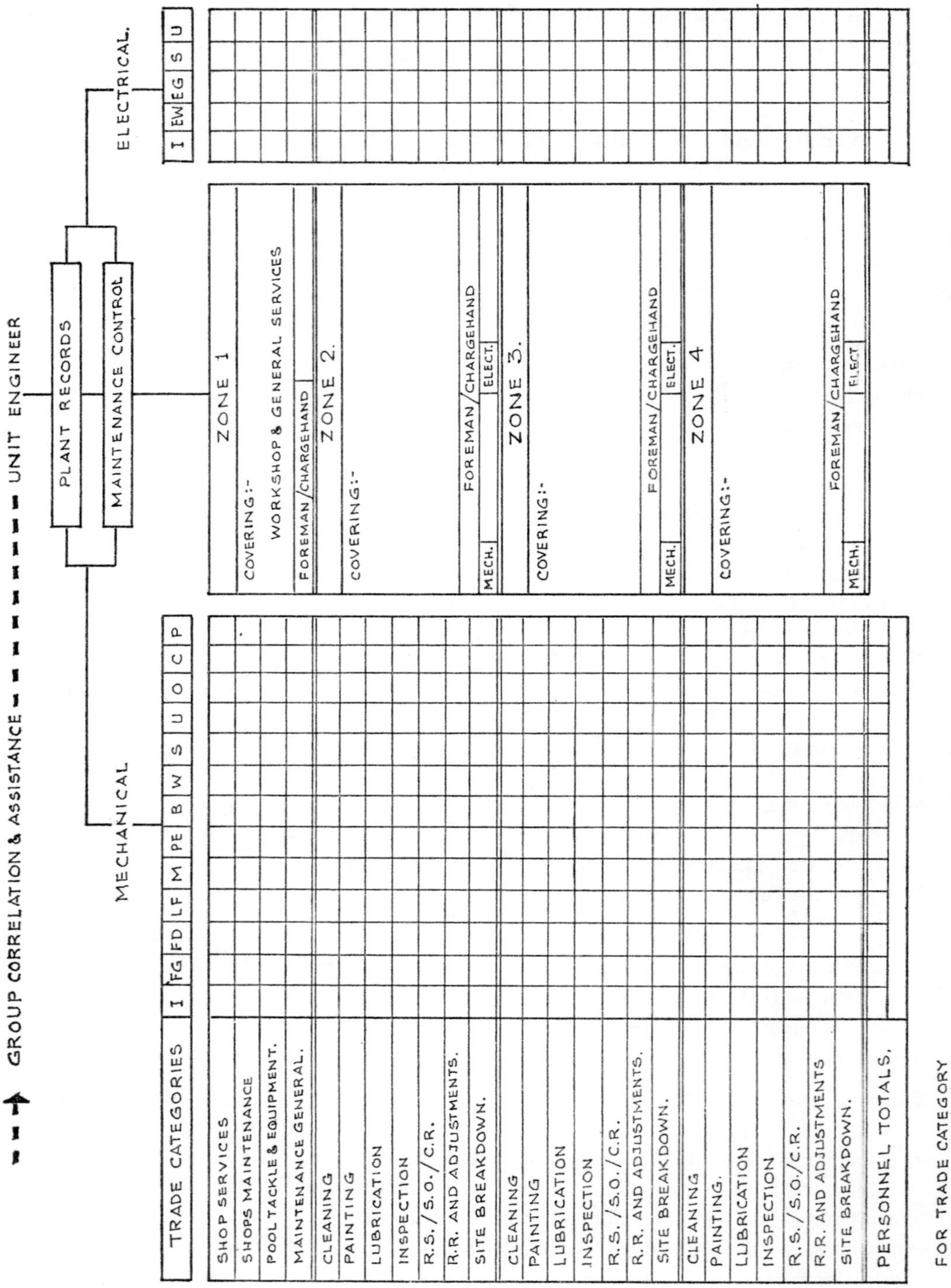


FIG. 5.2. Unit activity personnel and functional organisation chart (for trade category key see Fig. 4.4).

Adjusting the schedules to spread the manpower as evenly as possible, so that the minimum labour force is employed to the maximum purpose.
Allowance for unscheduled work.
The allocation of specific periods and times to carry out the jobs.
The provision of personnel to carry out the jobs.
The instruction and direction of personnel in their daily tasks.
The provision of supplies and materials to the job.
The transport of men and materials when required.
The recording of work done and outstanding.
The allocation of costs to the job.
The assessment of the operational facilities and services required, and their augmenting when necessary.
The maintenance accommodation required, and its layout.
Documentation, operational procedure, and control.

## THE ESTABLISHMENT OF THE BASIC PERSONNEL AND FUNCTIONAL ORGANISATION

From Fig. 5.1, the basic personnel and functional chart shown in Fig. 5.2 can be derived. It should be noted that the plant and equipment of the Unit have been divided into zones for service and maintenance purposes. To each zone is allocated a basic crew under the charge of the Unit's mechanical and electrical engineers, with a foreman or leading hand to oversee the various elements of maintenance within the zone. The zones are supported by the Unit's workshops and general services (designated as zone 1), and reinforcing these in turn, through the engineers-in-charge, is the Group's engineering organisation. Work scheduling and control is effected, and records are maintained, through the Unit's plant records and maintenance control centres, which cooperate with their counterparts at Group level. The chart gives a composite picture of the general organisation of personnel; its deployment and occupational categories, in the various spheres of maintenance and zones of activity; and the services supporting the 'field' zones. The numbers and deployment of personnel are derived by considering the plant and equipment involved, its type and location, its maintenance demands over specific periods, and the availability of labour and plant.

## ASSESSMENT OF THE UNIT'S ENGINEERING WORK LOAD

To simplify future scheduling and control, it is convenient to regard the Unit's engineering work load as the total of the work loads in the various zones already defined. Attention may therefore be confined to one typical zone.

Site layout drawings and flow diagrams (where appropriate) should be prepared, illustrating the plant arrangement in the sections of the zone, and identifying each item of

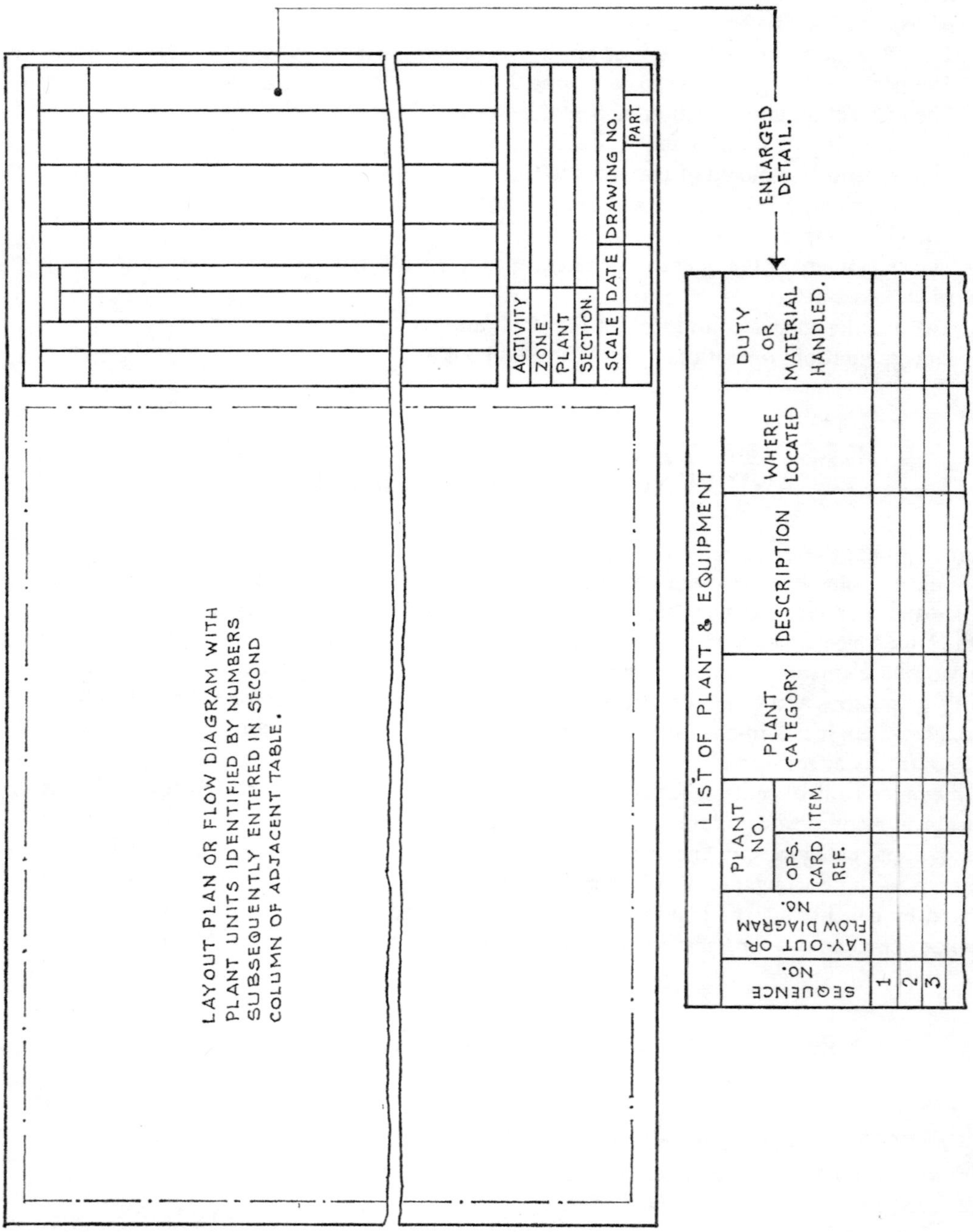


FIG. 5.3. Lay-out or flow diagram with plant identity.

| UNIT ACTIVITY | | PLANNED MAINTENANCE PLANT SUMMARY SHEET. |
|---|---|---|
| MAINTENANCE ZONE | | |

| PLANT CATEGORY | OPS. CARD REF. | NUMBER TO COVER | UNIT ANNUAL MAN-HOUR DEMAND. | ANNUAL MAN HR. DEMAND | REMARKS. |
|---|---|---|---|---|---|
| | | | | | |
| | | | | | |
| | | | | | |
| | | | | | |
| | | | | | |
| | | | | | |
| | | | | | |
| | | | | | |
| | | | | | |
| | | | | | |
| | | | | | |
| | | | | | |

| TOTAL ANNUAL MAN-HOUR DEMAND (ON SCHEDULED WORK) | | NO. OF MEN BASED ON 50 WEEKS OF | | |
|---|---|---|---|---|
| UNSCHEDULED WORK ALLOWANCE ----- ☐ % = | | 40 HRS. | 48 HRS. | 56 HRS. |
| ABSENTEEISM & HOLIDAYS ALLOWANCE -- ☐ % = | | | | |
| TRAVEL TIME, COLLECTING TOOLS & PARTS -- ☐ % = | | | | |
| WAITING TIME ---------------- ☐ % = | | | | |
| MAKING-UP REPORTS ------------ ☐ % = | | | | |
| TOTAL ESTIMATED MAN-HOURS: | | | | |

| OCCUPATIONAL CATEGORY | | CODE | |
|---|---|---|---|
| | | | |

FIG. 5.4. Plant summary sheet.

plant, together with its location and category. The necessary data can be tabulated on each drawing as shown in Fig. 5.3.

From the foregoing, plant summary sheets are prepared as shown in Fig. 5.4, listing the main categories of plant in alphabetical order of their plant number letters, and indicating the numbers in each category, with an operations card reference number in each case. The latter comprises the first two portions of the plant number.

Maintenance operations schedules are next prepared for each category and type of plant shown in the plant summary sheets, giving references to essential data, and listing the operations to be carried out; the maintenance element applicable, as a guide to the occupational category of the men involved; the frequency of the operations; and the manning and hours needed in each occupational category.

A typical maintenance operations schedule prepared on this basis is shown in Fig. 5.5. They should be drawn up by combining experience with time study where appropriate, and may be adjusted as further experience dictates. In this process, the makers' constructional and assembly drawings and recommended maintenance routines should be studied, but adjusted if necessary to avoid over-elaboration of the total work load finally created. The application of network techniques in the planning of maintenance operations, to ensure efficient sequencing of tasks and deployment of manpower is also relevant here (*see* Appendix II). Each maintenance operations schedule should be filed vertically in alphabetical order of their plant record number codings, with signals to facilitate ready location, withdrawal, and replacement; for example, 1/CON indicates the first category of conveyor, and applies to all conveyors of that category in the zone, Unit, and Group; each has its separate plant number, such as 1/CON/03, for purposes of identification.

From the maintenance operations schedules described, and in readiness for preparing the master schedules, maintenance patterns can be drawn up in the form shown in Fig. 5.6, for each plant category. These show the phasing of the maintenance operations on an annual and longer term basis, under each element of maintenance and occupational category, together with an estimate of the man-hours and labour cost involved. Interdependent operations are enclosed within squares; any change in timing in one of these involves a like change in the others. Circles denote independent operations, which can be separately adjusted in the scheduling process. The use of this feature makes it easier to achieve satisfactory work-load spreading in making up the master schedules. Using the data from the maintenance patterns, the next step is to complete the plant summary sheets (Fig. 5.4) for each occupational category, by entering the man-hour demands for each type of plant, together with the appropriate allowances, to obtain an assessment of manpower requirements in the zone.

At this stage, in preparation for the scheduling process, the element man-hour loading charts should be prepared and filed in correct element sequence, in a suitable folder (Fig. 5.7). In these, the man-hour loading for the particular element of maintenance covered is built up vertically for each week of the year. The appropriate man-hour scale is attached at the left of the chart, and is graduated on the basis of the total man-hours envisaged in any week, obtained from the plant summary sheets already referred to. The appropriate allowance scaling for the particular maintenance element and occupational category is attached at the right side of the chart.

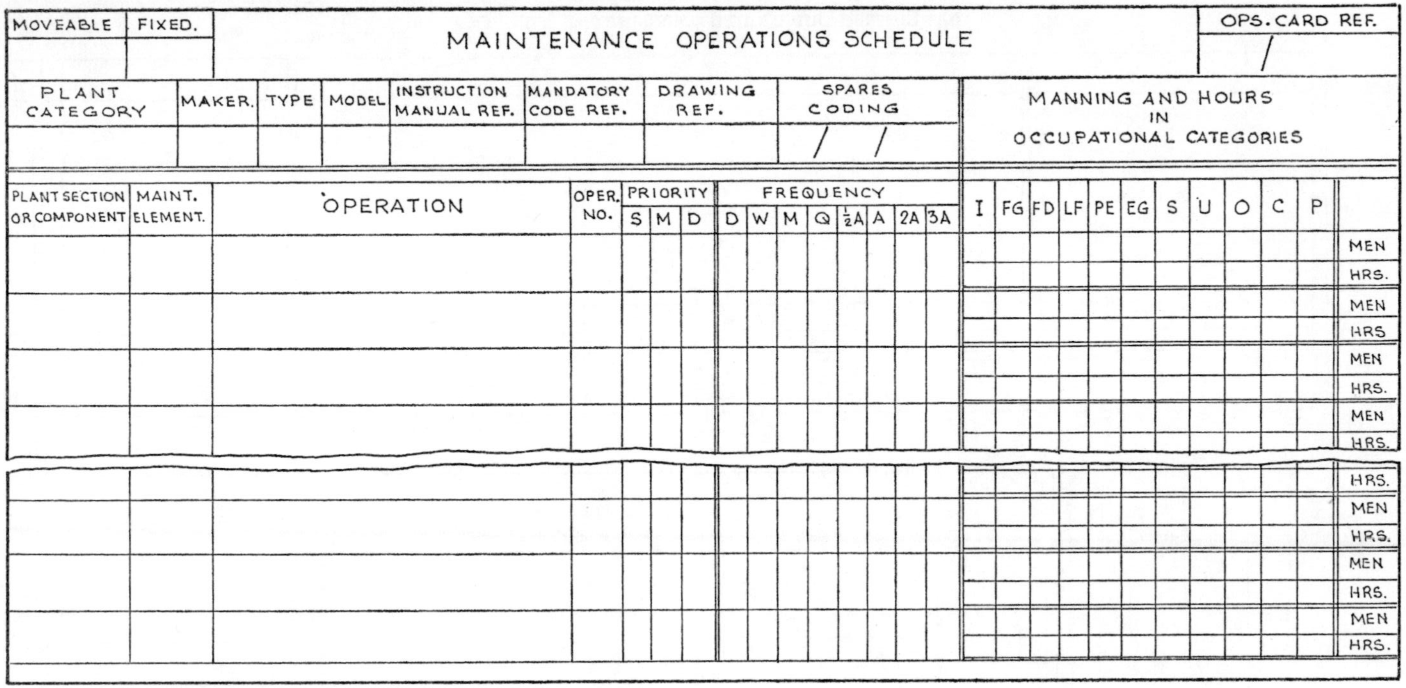


CODINGS:

| MAINTENANCE ELEMENTS | |
|---|---|
| C | CLEANING |
| P | PAINTING |
| L | LUBRICATION |
| I | INSPECTION |
| RS. | ROUTINE SERVICING |
| S.O. | SITE OVERHAUL. |
| C.R. | CYCLIC REPLACEMENT. |
| RR&A | RUNNING ADJUSTMENTS. |

| PRIORITY | |
|---|---|
| S | STATUTORY |
| M | MANDATORY |
| D | DESIRABLE. |

| FREQUENCY | |
|---|---|
| D | DAILY |
| W | WEEKLY |
| M | MONTHLY |
| Q | QUARTERLY |
| ½ A | HALF-ANNUAL. |
| A | ANNUAL |
| 2 A | EVERY TWO YEARS. |
| 3 A | EVERY THREE YEARS. |

| FREQUENCY ENTRIES | |
|---|---|
| ○ | GREEN - INDEPENDENT OF OTHER ELEMENTS. |
| ○ | YELLOW - REPEATED IN LONGER TERM OPERATIONS. |
| □ | RED - TIED UP WITH OTHER ELEMENTS |
| □ | YELLOW - REPEATED IN LONGER TERM OPERATIONS |

| OCCUPATIONAL. |
|---|
| SEE FIG. 4:4. |

FIG. 5.5. Maintenance operations schedule. Codings: maintenance elements, Fig. 5.5; crafts, Fig. 4.4; frequencies, Fig. 5.5.

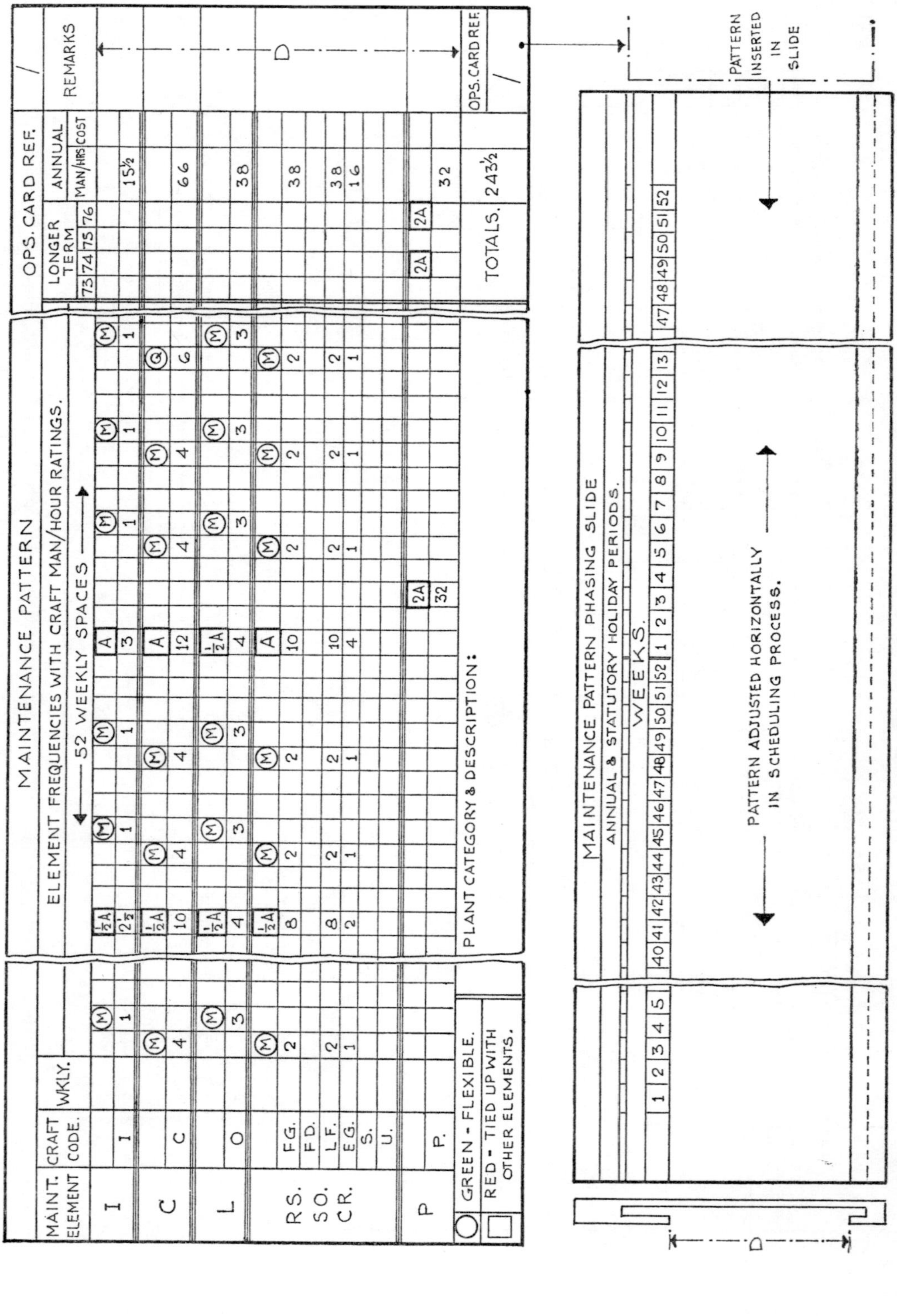


FIG. 5.6. Maintenance pattern and phasing slide.

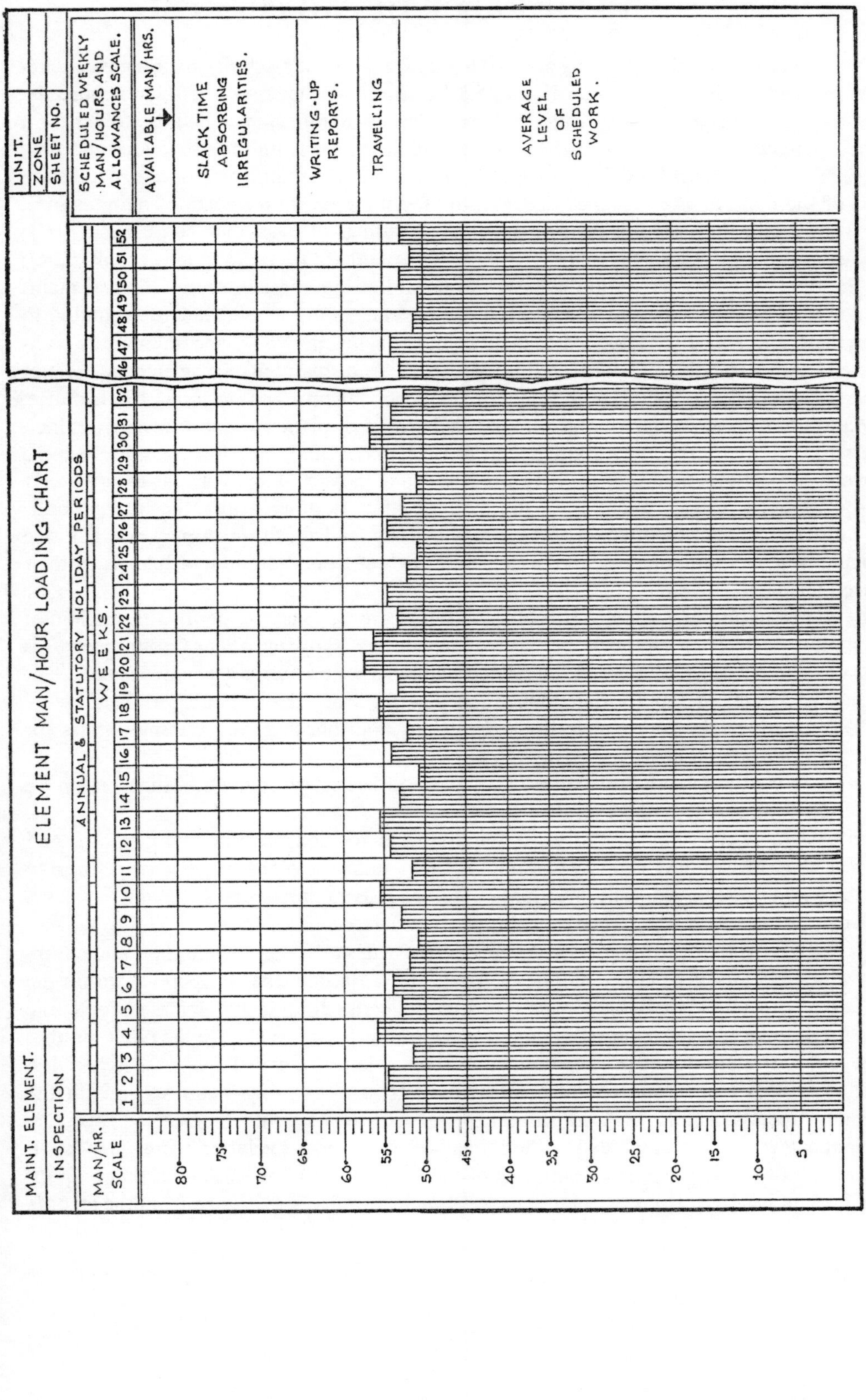


FIG. 5.7. Element man/hour loading chart. *Note:* Man/hour and allowance scales designed and attached to suit the maintenance element being dealt with.

## SCHEDULING OF THE WORK

With the data so far collated, it is possible to prepare the Unit's activity maintenance schedules for each maintenance zone. Figure 5.8 illustrates the make-up of such a schedule. The base page represents a leaf from the Unit's activity maintenance schedules folder for a particular maintenance zone. At the left are entered the data from the site layout and flow diagrams already referred to. To the right of this plant and equipment schedule, details are given of the operations, frequency and phasing for the relevant elements of maintenance. These latter schedules are sub-divided under the headings of Inspection (I); Cleaning (C); Lubrication (L); Routine Servicing (RS)/Site Overhaul (SO)/Cyclic Replacement (CR); and Painting (P). This separation is based on the following considerations. First, if maintenance is to be thorough and effective, all of its elements must be included. Secondly, to cover these in a single schedule is impracticable; the schedule would become unduly complicated, and also the frequency coding of the separate maintenance elements are not necessarily phased to coincide, even when the frequencies themselves are equal. For instance, the quarterly inspection operations on a particular item of plant are not necessarily coincident with the quarterly lubrication operations on that same plant item. Similarly some quarterly inspection operations may require a machine to be running, whereas the quarterly routine servicing is likely to call for a standing machine. In other words, the segregation is into those elements of maintenance which may overlap, and those which may not. Routine servicing, site overhaul, and cyclic replacement are grouped under one schedule, as by their nature they cannot overlap.

The co-ordination of work and allocation of duties are also made easier by this method of segregation. It helps to establish the basis of scheduling, showing where Group control is involved. The reference number of the related maintenance operations schedule is included; this in turn locates the appropriate maintenance pattern for the plant category being scheduled, from which the coded operations, frequency, and phasing, and the man-hours for each trade can be obtained.

A note is included of the incidence of annual and statutory holidays, which may have to be used or avoided, as the case may be.

The system also establishes forward requirements for longer-term maintenance, and sets out the prearranged running periods of those items of plant where change-over to standby plant is involved. The information relating to maintenance scheduling of such items can be entered in the 'remarks' column.

It is reasonable to regard the work as being concerned, in the main, with plant and equipment which may itself be either mobile or fixed, but is static in that it operates within the confines of a single maintenance zone. On this understanding, a unit released for cyclic overhaul is replaced by its equivalent, and it is merely necessary to change the plant number in the appropriate column. Redundant or obsolete plant is deleted from the schedules. Additional plant can be incorporated into gaps left in the various plant sections in the schedules. Items of plant which are subject to frequent re-siting may be grouped together; the plant category is then regarded as the primary classification, the latter being amended in the schedules to avoid constant reshuffling of the whole scheduling of this group of equipment. If plant is moved from one maintenance zone to another, it is deleted from the records

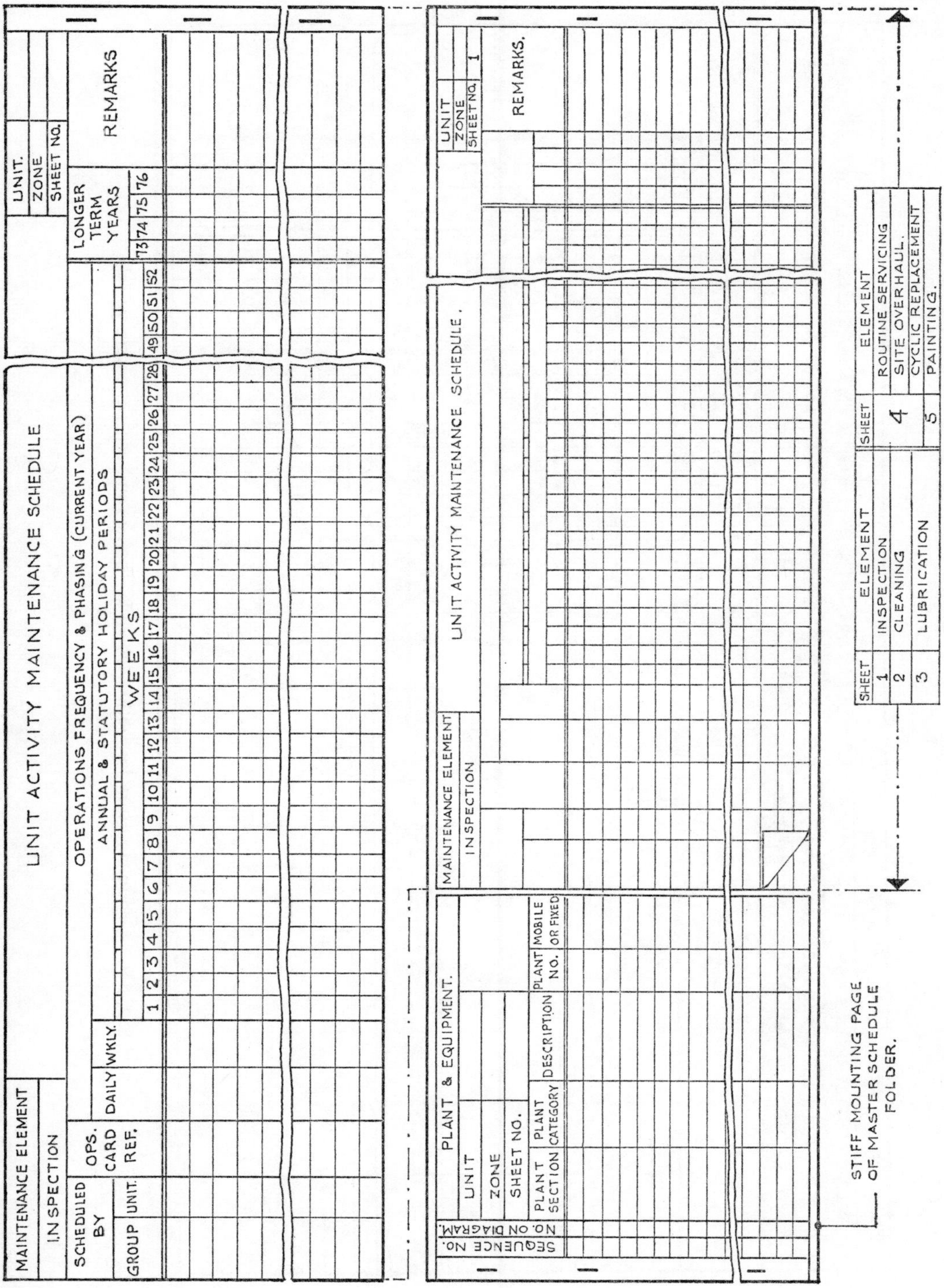


FIG. 5.8. Unit activity maintenance schedule.

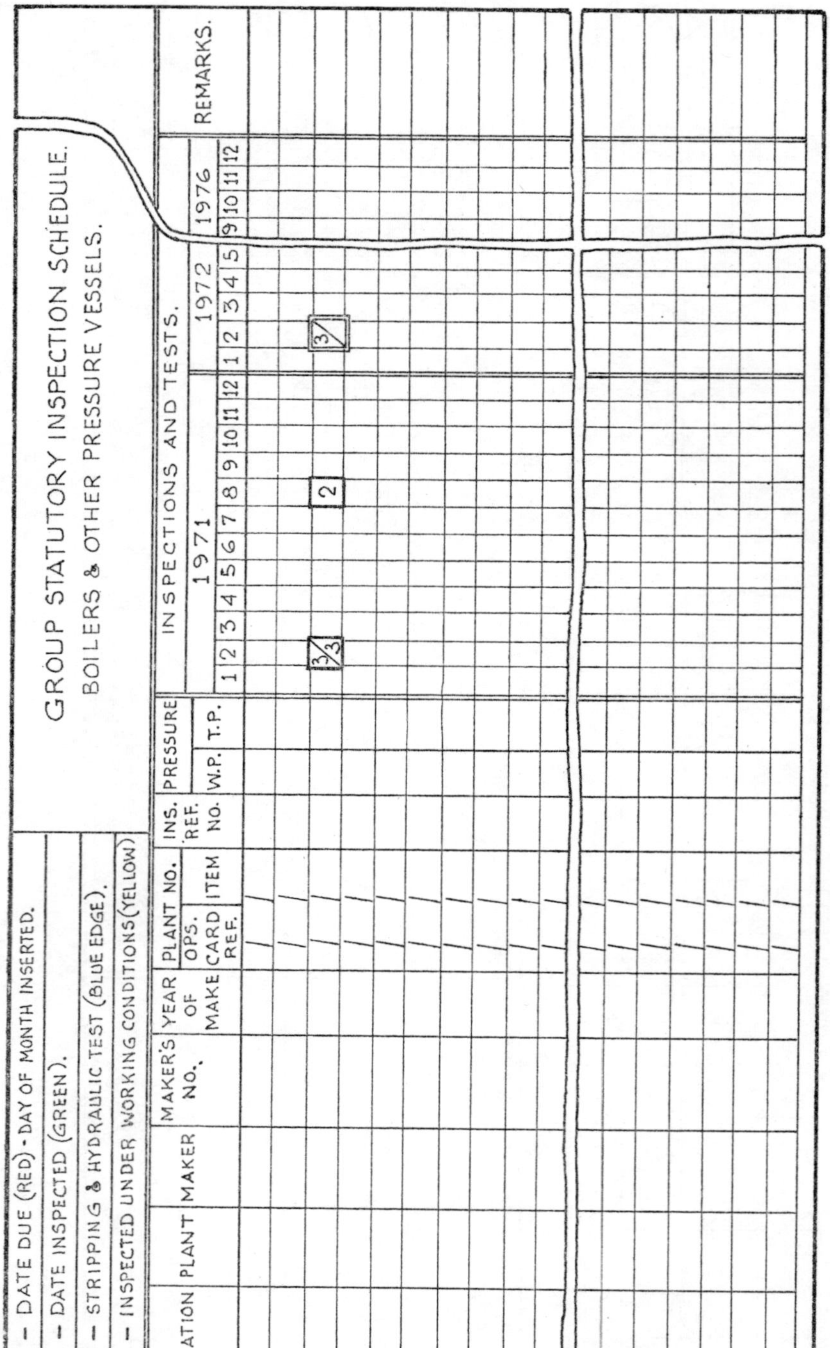


FIG. 5.9. Group statutory inspection schedule—boilers and other pressure vessels.

**GROUP MONTHLY SCHEDULE**
**STATUTORY INSPECTIONS.**
**STEAM BOILERS & OTHER PRESSURE VESSELS.**
FOR MONTH OF..................19.......

FROM: | TO:

| LOCATION & ITEM. | PLANT NO. | INS. REF. NO. | WEEK ALLOCATED | | | | DATE FIXED BY INS. CO. |
|---|---|---|---|---|---|---|---|
| | | | 1 | 2 | 3 | 4 | |
| | | | | | | | |
| | | | | | | | |

TO BE RETURNED BY INSURANCE CO. TO GROUP ENGINEER'S OFFICE CONFIRMING DATES IN THE LAST COLUMN.

| FOR GROUP OFFICE | DATE | FOR INSURANCE CO. | DATE. |
|---|---|---|---|
| | | | |

ISSUED BY GROUP ENG. OFFICE TO INSURANCE CO. IN MONTH PRECEDING. DUPLICATE RETAINED. INSURANCE CO. RETURN CONFIRMING DATES.

**GROUP MONTHLY ADVICE**
**STATUTORY INSPECTIONS**
**STEAM BOILERS & OTHER PRESSURE VESSELS.**
FOR MONTH OF..............19.....

FROM: | TO:

PLEASE MAKE THE NECESSARY ARRANGEMENTS TO MEET THE INSPECTION REQUIREMENTS BELOW:-

| PLANT ITEM | PLANT NO. | INS. REF NO. | DATE FIXED | REQUIRED | | | REMARKS. |
|---|---|---|---|---|---|---|---|
| | | | | INT. EXAM. | STRIP COVER | HYD. TEST. | |
| | | | | | | | |
| | | | | | | | |

ANY ALTERATION TO DATE FIXED TO BE NOTIFIED TO GROUP ENG. OFFICE FOR CONFIRMING WITH THE INSURANCE CO.
IF VERBAL REPORT LEFT BY INSPECTOR REQUIRES REPAIRS OUTWITH UNIT RESOURCES, NOTIFY GROUP ENG. OFFICE.
DO NOT AWAIT OFFICIAL REPORT FORM.

| FOR GROUP OFFICE | DATE |
|---|---|
| SIGNED:.................................... | .................... |

ISSUED BY GROUP ENG. OFFICE TO UNIT ACTIVITY ENGINEER IN TIME TO MAKE PREPARATIONS. DUPLICATE RETAINED.

**REPAIRS NOTIFICATION & COMPLETION ADVICE**
**ARISING FROM STATUTORY INSPECTIONS.**
**STEAM BOILERS & OTHER PRESSURE VESSELS**

FROM: | TO:

| DATE INSPECTED | INS. REPORT DATE (ATTACH) | PLANT ITEM | PLANT NO. | INS. REF. NO. | REMARKS |
|---|---|---|---|---|---|
| | | | | | |

MATTERS FOR ATTENTION:

PERMISSIBLE WORKING PRESSURE FOR NEXT 14 MONTHS AFTER REPAIRS COMPLETE [ ]

SIGNED:............................... (GROUP OFFICE) DATE:................

DETACH & RETURN ON COMPLETION OF REPAIRS.

| INSPECTION REPORT DATE | INSPECTION DATE | PLANT ITEM. | PLANT NO. | INS. REF. NO. | REMARKS |
|---|---|---|---|---|---|
| | | | | | |

REPAIRS ADVISED.................... COMPLETED.........................

SIGNED........................(UNIT ENGINEER) DATE ...............

TO BE ATTACHED TO UNIT COPY OF OFFICIAL REPORT SENT BY GROUP OFFICE TO UNIT ENGINEER. DUPLICATE COPY RETAINED WITH COPY REPORT.

FIG. 5.10. Statutory inspections (boilers and pressure vessels)—associated documentation.

of the former zone, and added in its new zone. In some cases, it may be helpful to provide a method of inserting strips in these columns to permit easier modification.

Before considering the actual scheduling, it is appropriate to consider the procedure for the scheduling of the Group-controlled maintenance at a Unit. Referring back to the categories of work which should be scheduled and controlled on a Group basis, the following are taken as examples for the present purpose.

### *Boilers, superheaters, economisers, steam receivers, air receivers, and other pressure vessels*

Figures 5.9 and 5.10 illustrate a method of controlling, on a Group basis, the registration, certification, inspection, and maintenance of plant in these categories. From the chart shown, a general picture of monthly commitments is obtained and can be incorporated into the Unit's maintenance schedules; the actual weeks and days are confirmed monthly in advance, to allow for contingencies. Arrangements are incorporated to ensure effective Group control, and also that the validity of insurance certificates is maintained. A schedule chart of the type indicated, 760 mm wide × 610 mm deep, with about 6½ mm squares per month per item, will suffice for about 80 items of plant for six years, with forward and back reference. The chart forms the basis of control; and by linking-up this forward phasing with the insurance company's commitments, mutual benefit is derived and smoother working assured. The plant record and inventory section up-dates the technical and other data for all items, as a necessary adjunct to these maintenance control processes.

### *Cranes and lifting tackle*

The arrangements for Group control over the registration, specifying, ordering, certification, inspection, maintenance, and replacement of plant in this category can be similar to that in the foregoing section. There are differences in detail and procedure arising from the different nature of the equipment, and the fact that it comprises both fixed and transportable plant items. Following a survey of all equipment in the Group, the necessary statutory registers are prepared in duplicate, a full set being retained at the Group's office, and each Unit being furnished with its own copy for reference by the inspectorate; these must be kept up to date. The general form of the sheets in the register is shown in Fig. 5.12. All Units should then be issued with a directive of procedure.

Figures 5.11 to 5.15 illustrate this method of establishing and maintaining control on a Group basis.

## Visits by inspectors

This heading includes all work in connection with visits of the Government inspectors appointed to ensure the fulfilment of current regulations related to safety, health, hygiene, and conditions of employment. The responsibilities laid upon employers in this connection make it essential for strict control to be exercised by the Group to ensure compliance with the regulations, and also that when the inspectors are dissatisfied, the matters are promptly and satisfactorily resolved. Regulations are drawn up in the interests of all, but cannot

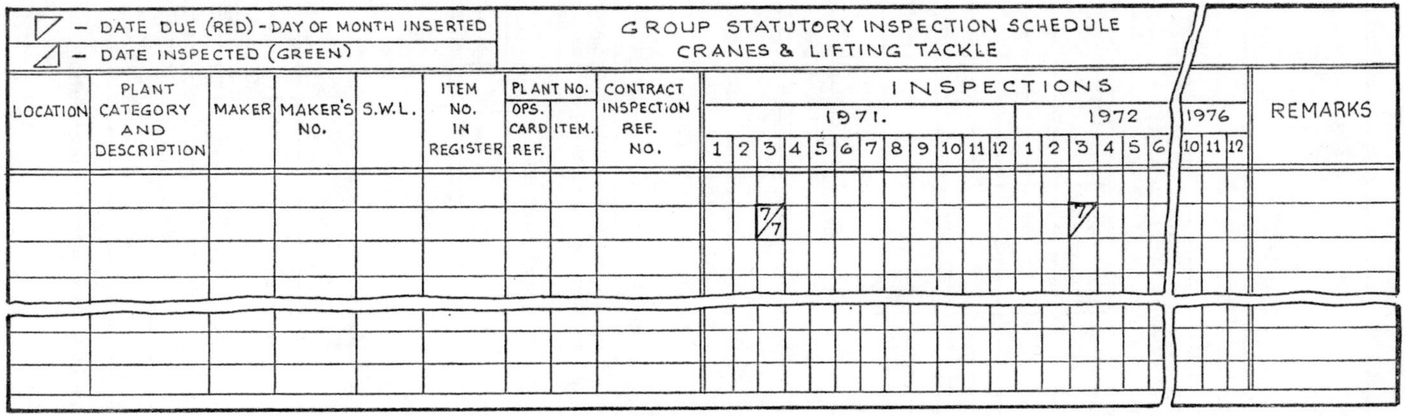


FIG. 5.11. Group statutory inspection schedule—register of cranes and lifting tackle.

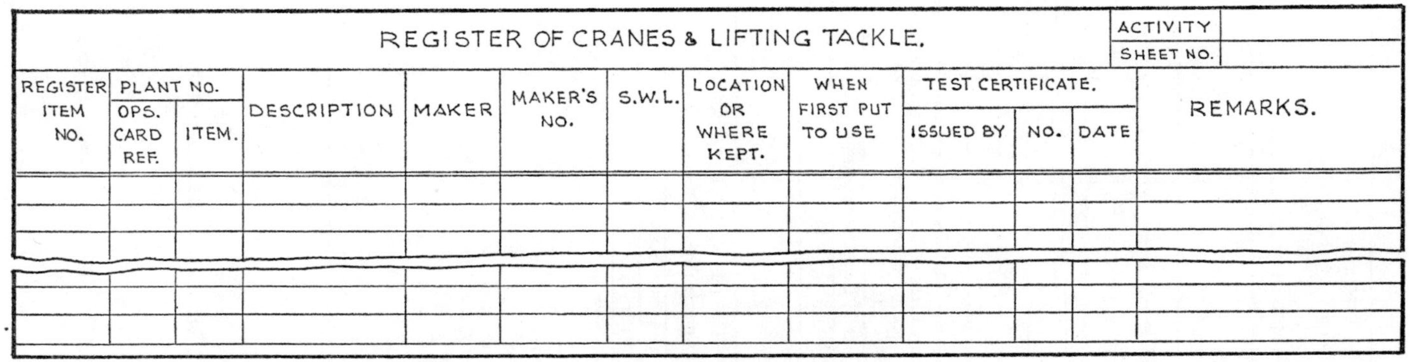


FIG. 5.12. Register of cranes and lifting tackle.

**GROUP MONTHLY SCHEDULE**
**STATUTORY INSPECTIONS.**
**CRANES & LIFTING TACKLE.**
FOR MONTH OF............ 19.....

| LOCATION AND ITEM. | ITEM NO. IN REGISTER. | CONTRACT INSPECTION REF. NO. | LAST INSPECTION DATE | DATE FIXED BY INSP$^{N.}$ CO. |
|---|---|---|---|---|
| | | | | |
| | | | | |
| | | | | |
| | | | | |
| | | | | |
| | | | | |
| | | | | |

THE INSPECTION UNDERTAKING ARE ASKED TO RETURN THIS TO THE GROUP ENGINEERING OFFICE WITH CONFIRMING DATES IN THE LAST COLUMN.

| FOR GROUP OFFICE | DATE | FOR INSPECTION CO. | DATE |
|---|---|---|---|
| | | | |

**MONTHLY ADVICE**
**STATUTORY INSPECTIONS.**
**CRANES & LIFTING TACKLE.**
FOR MONTH OF............19...

| TO | |
|---|---|
| ENGINEER-IN-CHARGE | |
| UNIT ACTIVITY | |

THE FOLLOWING CONTRACT INSPECTIONS ARE ARRANGED FOR THE DATE SHOWN. PLEASE ARRANGE TO HAVE THE PLANT & EQUIPMENT AVAILABLE:-

| DESCRIPTION OF ITEM | ITEM NO. IN REGISTER. | INSPECTION DATE. | REMARKS |
|---|---|---|---|
| | | | |
| | | | |
| | | | |
| | | | |
| | | | |
| | | | |

SIGNED.............. (GROUP OFFICE) DATE..........

**STATUTORY INSPECTION REPORTS.**
**CRANES & LIFTING TACKLE.**

| TO | |
|---|---|
| ENGINEER-IN-CHARGE | |
| UNIT ACTIVITY | |

WE ENCLOSE COPIES OF REPORTS OF INSPECTIONS CARRIED OUT ON................(DATE). PLEASE GIVE ATTENTION TO ANY REMARKS MADE AND EFFECT ANY REPAIRS, NOTIFYING COMPLETION TO THIS OFFICE IN THE APPROPRIATE SPACE ON THE SLIP HEREWITH.

IF THERE IS ANY WORK WHICH CANNOT BE UNDERTAKEN BY YOUR UNIT, THIS SHOULD BE NOTIFIED TO THIS OFFICE FOR ATTENTION, IN THE APPROPRIATE SPACE ON THE SLIP ATTACHED.

SIGNED.....................(GROUP OFFICE) DATE...............

**STATUTORY INSPECTIONS.**
**CRANES & LIFTING TACKLE.**

TO:
GROUP ENGINEERING OFFICE.

THE REMARKS AND REPAIRS ON THE REPORTS SHOWN HAVE NOW BEEN ATTENDED TO :-

| ITEM NO. | DESCRIPTION. | INSPECTION DATE FROM LAST REPORT. |
|---|---|---|
| | | |
| | | |
| | | |

THE REQUIREMENTS BELOW, CALLED FOR IN THE REPORTS INDICATED, CANNOT BE UNDERTAKEN AT THE UNIT ACTIVITY:-

| ITEM NO. | DESCRIPTION | INSPECTION DATE | REQUIREMENTS. |
|---|---|---|---|
| | | | |
| | | | |
| | | | |
| | | | |

| ACTIVITY | SIGNED..................(UNIT ENGINEER) |
|---|---|
| | DATE.................19...... |

NOTE:
DUPLICATE RETAINED AT SOURCE IN EACH CASE.

FIG. 5.13. Statutory inspection (cranes and lifting tackle)—associated documentation.

LIFTING TACKLE REQUIREMENTS SLIP

| FROM | UNIT ACTIVITY | AUTHORISED BY | DATE | TO | GROUP ENGINEERING OFFICE MAINTENANCE CONTROL. | RECEIVED BY | DATE |
|---|---|---|---|---|---|---|---|
| | | | | | | | |

THE FOLLOWING TACKLE, IS REQUIRED AT THIS UNIT TO REPLACE ITEM NO. / FOR ADDITION TO THAT ALREADY COVERED UNDER THE INSPECTION AND CERTIFICATION ARRANGEMENTS:-

CHAIN BLOCKS:

| NO. REQD. | CAPACITY | TYPE | HOOK TO HOOK LENGTH (EXTENDED) | IS BALL SWIVEL HOOK REQD. | REMARKS |
|---|---|---|---|---|---|
| | | | | | |
| | | | | | |
| | | | | | |

ROPE BLOCKS:

| NO. REQD. | CAPACITY | NO. OF SHEAVES | ROPE DIAMETER | HEIGHT OF LIFT REQUIRED | REMARKS. |
|---|---|---|---|---|---|
| | | | | | |
| | | | | | |
| | | | | | |

LIFTING CHAINS:

| NO. REQD. | S.W.L. | NO. OF LEGS | LENGTH OF LEG | SIZE OF CHAIN | TOP RING SIZE | REMARKS |
|---|---|---|---|---|---|---|
| | | | | | | |
| | | | | | | |
| | | | | | | |

WIRE OR ROPE SLINGS:

| NO. REQD. | S.W.L. | NO. OF LEGS | LENGTH OF LEG | DESCRIPTION OF ENDS | TYPE | SIZE OF ROPE | REMARKS |
|---|---|---|---|---|---|---|---|
| | | | | | | | |
| | | | | | | | |
| | | | | | | | |

SHACKLES, HOOKS & SWIVELS (ATTACH SKETCH OR GIVE SKETCH NO.):

| NO. REQD. | DESCRIPTION. | S.W.L. | REMARKS. |
|---|---|---|---|
| | | | |
| | | | |
| | | | |

FIG. 5.14. Lifting tackle requirements slip.

## LIFTING TACKLE RECEIVING SLIP.

| FROM | UNIT ACTIVITY. | ISSUED BY | DATE | TO | GROUP ENGINEERING OFFICE | RECEIVED BY | DATE |
|---|---|---|---|---|---|---|---|
| | | | | | MAINTENANCE CONTROL | | |

THE FOLLOWING TACKLE HAS BEEN RECEIVED AT THIS UNIT FOR INCLUSION UNDER THE INSPECTION AND CERTIFICATION ARRANGEMENTS:

CHAIN BLOCKS:

| MAKER. | DATE RECD. | ORDER NO. | CAPACITY | HK. TO HK. LENGTH | MAKER'S NO. | ADDITIONAL | REPLACEMENT. | REG. NO. | OLD ITEM SCRAPD. |
|---|---|---|---|---|---|---|---|---|---|
| | | | | | | | | | |
| | | | | | | | | | |
| | | | | | | | | | |

ROPE BLOCKS:

| MAKER. | DATE RECD. | ORDER NO. | CAPACITY | NO. OF SHEAVES | | ROPE SIZE | MAKERS NO. | ADDITIONAL | REPLACEMENT. | REG. NO. | OLD SET SCRAPPED. |
|---|---|---|---|---|---|---|---|---|---|---|---|
| | | | | | | | | | | | |
| | | | | | | | | | | | |
| | | | | | | | | | | | |

LIFTING CHAINS:

| MAKER | DATE RECD. | ORDER NO. | S.W.L. | NO. OF LEGS | CHAIN SIZE | RING SIZE | LENGTH | ADDITIONAL | REPLACEMENT | REG. NO. | OLD ITEM SCRAPPED |
|---|---|---|---|---|---|---|---|---|---|---|---|
| | | | | | | | | | | | |
| | | | | | | | | | | | |
| | | | | | | | | | | | |

WIRE OR ROPE SLINGS:

| MAKER | DATE RECD. | ORDER NO. | S.W.L. | TYPE | ROPE SIZE | LENGTH | ADDITIONAL | REPLACEMENT. | REG. NO. | OLD ITEM SCRAPPED |
|---|---|---|---|---|---|---|---|---|---|---|
| | | | | | | | | | | |
| | | | | | | | | | | |
| | | | | | | | | | | |

SHACKLES, HOOKS, AND SWIVELS. (SKETCH TO BE ATTACHED OR SKETCH NO. GIVEN):

| MAKER | DATE RECD. | ORDER NO. | S.W.L. | DESCRIPTION | ADDITIONAL | REPLACEMENT. | REG. NO. | OLD ITEM SCRAPPED |
|---|---|---|---|---|---|---|---|---|
| | | | | | | | | |
| | | | | | | | | |
| | | | | | | | | |

FIG. 5.15. Lifting tackle receiving slip.

cover all contingencies, nor be tailored to fit every case. They must be interpreted sensibly and it is important that mutual confidence is established between the inspectors, and the firm or undertaking. Internal standards should be higher than the minimum called for under the regulations, where merited, and every effort must be made to attain and maintain them, with effective control. Evidence of this, and the timely calling in of the inspectors for discussion, can do much to establish reasonable interpretation and implementation of the regulations, and the mutual confidence referred to.

The Government's inspectors are empowered to visit a firm's premises at any time. Any subsequent reports submitted by the inspectors should be directed in duplicate, via the secretariat, to the Group's engineering office. Any action required should be notified in writing to the Unit, together with one copy of the inspector's report. Completion of the action should be noted on the reports at both offices, and the Group's engineering office should notify the inspector, through the secretariat, as quickly as possible. The reports should be filed for each Unit, in sequence of the dates of inspection.

## Major overhauls of fixed plant at site

This aspect is illustrated by considering one type of plant (a steam turbine with auxiliaries), and establishing a suitable procedure and control for it. Figures 5.16 and 5.17 show how the procedure and control operate. It is important to insist that evidence of a high standard of inspection and overhaul is provided, in the shape of thorough and well-presented reports, giving a chronological record and history of the performance and maintenance of the plant during its life.

## Periodic replacement of plant units

From the Unit's standpoint, the work is resolved into an element of maintenance which has been referred to as 'Periodic replacement of mobile or transportable plant'. To provide for this category of work in the Unit's maintenance schedules, the Group should provide to the various Units an annual summary of plant items due for replacement, giving for each its location, identity, the week it is due at the Group's workshops, and the week the plant unit to replace it is expected to arrive at the Unit (Fig. 5.18). Under the procedures for the Group's plant record and central workshops, a time interval between the receipt of the replacement plant at the Unit and the arrival of the released plant at the Group's workshops has already been agreed. This is essential to maintain the turn-round of periodic overhauls at the workshops. However, to provide flexibility and allow for contingencies, the actual periodic replacement programme should be confirmed monthly in advance to the Units, so that they can confirm or adjust their own weekly work schedules accordingly. The monthly advice note serves to give the up-to-date details of the type and identity of the replacement plant being sent, its date of dispatch, the location and identity of the plant it is to replace, and the date the released plant is due at the central workshops (Fig. 5.19). Such advice supersedes the information in the annual summary. As experience is gained, and the arrangements become more closely attuned to the overall operational activity the divergence of the forward monthly programme from the annual forecast should become progressively smaller.

| 10-15 | SCHEDULED PERIOD IN MONTH FOR INSPECTION UNDER RUNNING CONDITIONS AND CHECKING OF SPARES POSITION BY MAKERS' ENGINEER (YELLOW) |
|---|---|
| (blank box) | —SIGNIFIES DONE (GREEN). |
| 8-22 | —SCHEDULED PERIOD IN MONTH FOR MAJOR BI-ANNUAL OVERHAUL (RED) |
| 8-20 | —ACTUAL PERIOD OF OVERHAUL (GREEN) |

**GROUP MAJOR OVERHAUL SCHEDULE.**
**STEAM TURBINES & AUXILIARY PLANT**

| LOCATION | MACHINE | MAKER | MAKER'S No. | TYPE | SIZE OR DUTY (KW) | DRIVE | PLANT No. | OPS. CARD REF. | LAST OVERHAUL PRIOR TO SCHEDULE | OVERHAUL PHASING 1971 1 | 2 | 3 | 4 | 5 | 6 | 7 | 8 | 1973 7 | 8 | 9 | 10 | 11 | 12 | REMARKS |
|---|---|---|---|---|---|---|---|---|---|---|---|---|---|---|---|---|---|---|---|---|---|---|---|---|
| | | | | | | | | | | | | | | | | 10-15 | 8-22 / 8-20 | | | | | | | |

FIG. 5.16. Group major overhaul schedule—steam turbines and auxiliary plant.

ADDRESS IN DUPLICATE TO FIRM PROVIDING SERVICES REQUIRED.

ENTER UNDERTAKING'S GROUP ENGINEERING OFFICE ADDRESS.

STEAM TURBINES & AUXILIARY PLANT.
MAJOR OVERHAULS.

THE FOLLOWING PLANT IS SCHEDULED FOR OVERHAUL IN THE YEAR 19...... :-

| LOCATION | MACHINE | MAKER'S No. | PRE-OVERHAUL INSPECTION AND SPARES CHECK | | MAJOR OVERHAUL | |
|---|---|---|---|---|---|---|
| | | | PERIOD ALLOCATED | DATE AGREED | PERIOD ALLOCATED | PERIOD AGREED |

DELETE AS NECESSARY:
SERVICES OF SUPERVISING ENGINEER ONLY REQUIRED.
SKILLED LABOUR & SUPERVISION TO BE PROVIDED.
ALL LABOUR & SUPERVISION TO BE PROVIDED.

RETURN ONE SIGNED COPY. ORDER TO BE ISSUED THEREAFTER.

| SIGNED .................. (FOR GROUP ENGINEER) | SIGNED .................. FOR .............. (FIRM) |
|---|---|
| DATE ..............19...... | DATE ..............19....... |

FROM: GROUP MAINTENANCE CONTROL

TO: MAINTENANCE CONTROL UNIT ACTIVITY..............

STEAM TURBINES & AUXILIARY PLANT.
MAJOR OVERHAULS.19.....

THE FOLLOWING PERIODS & DATES ARE NOW AGREED FOR THE MAJOR OVERHAUL/S OF THE PLANT SHOWN:-

| MACHINE | MAKER | MAKER'S No. | PRE-OVERHAUL INSPECTION AND SPARES CHECK. | | OVERHAUL PERIOD | |
|---|---|---|---|---|---|---|
| | | | FROM | TO | FROM | TO |

MESSRS...................... ARE PROVIDING:-
SERVICES OF SUPERVISING ENGINEER ONLY.
ALL SKILLED LABOUR & SUPERVISION.
ALL LABOUR & SUPERVISION
} DELETE AS NECESSARY.

OFFICIAL ORDER PLACED. NO:..............DATED............

GROUP WORKSHOPS WILL PROVIDE..........................

SIGNED ........................ (FOR GROUP ENGINEER) DATE..........

FIG. 5.17. Group major overhaul schedule (steam turbines and auxiliary plant)—associated documentation.

ANNUAL SCHEDULE
CYCLIC REPLACEMENT OF PLANT UNITS.
YEAR :...........

| FROM | | | TO | | |
|---|---|---|---|---|---|
| GROUP ENGINEERING MAINTENANCE CONTROL | ISSUED BY | DATE | MAINTENANCE CONTROL UNIT ACTIVITY.............. | RECEIVED BY | DATE |

| PLANT CATEGORY. | MAKER. | MAKER'S NO. | PLANT NO. | | REPLACE^T. PLANT DUE AT SITE (WEEK NO.) | RELEASED PLANT DUE AT GROUP WORKSHOPS (WEEK NO.) | MAINTENANCE ZONE | | | |
|---|---|---|---|---|---|---|---|---|---|---|
| | | | MECH. | ELECT. | | | 1 | 2 | 3 | 4 |

FIG. 5.18. Annual schedule—periodic replacement of plant units.

MONTHLY FORWARD SCHEDULE.
CYCLIC REPLACEMENT OF PLANT UNITS.
MONTH OF............19...

| FROM | | | TO | | |
|---|---|---|---|---|---|
| GROUP ENGINEERING MAINTENANCE CONTROL | ISSUED BY | DATE | MAINTENANCE CONTROL UNIT ACTIVITY.............. | RECEIVED BY | DATE |

| PLANT CATEGORY | REPLACEMENT BEING ISSUED | | | | | PLANT TO BE RELEASED | | | | | MAINTENANCE ZONE | | | |
|---|---|---|---|---|---|---|---|---|---|---|---|---|---|---|
| | MAKER | MAKER'S NO. | PLANT NO. | | DUE AT SITE | MAKER | MAKER'S NO. | PLANT NO. | | DUE AT GROUP SHOPS | 1 | 2 | 3 | 4 |
| | | | MECH. | ELECT. | | | | MECH. | ELECT. | | | | | |

FIG. 5.19. Monthly forward schedule—periodic replacement of plant units.

## Work related to pool services of the group

These may include such equipment as a locomotive fleet, internal railway wagons, a fleet of road vehicles, or equipment such as mobile cranes, winches, lifting tackle, bull-dozers, and so on. While such equipment may be deployed at, or allocated to, the Unit, it must be controlled by the Group to ensure it is maintained and deployed to suit the best interests of the Group as a whole. Deployment, overhaul, and major servicing should therefore be Group functions. At the same time, the allocation of pool equipment to a Unit imposes certain maintenance and operational demands upon that Unit, which must be allowed for in its maintenance schedules.

Consider, for example, a fleet of locomotives. The deployment of the locomotives, and their overhaul, can be controlled under the arrangements for the periodic replacement of plant units, with the proviso that where movement over national lines is involved, the necessary arrangements must be made with British Rail and confirmed to the Unit concerned. The location and deployment of the locomotives can be recorded in the Group's plant records office, the forward overhaul programme being marked up on a three-year basis. A tally board is useful to give the overall picture of deployment at a glance. Complete mechanical overhauls, at the three-year intervals proposed, would be carried out at the central workshops. Secondary annual overhauls may be done at the central workshops or at the Unit, as conditions dictate. The mechanical overhauls scheduled by the Group, and regular running maintenance scheduled by the Unit, should be incorporated in the maintenance schedules of the Unit employing the locomotive.

As redeployment takes place, the Group's maintenance control must clear with the Units concerned the current phasing of the inspections and overhauls relating to any locomotive allocated to them. It must also ensure that the necessary reports are transferred with the locomotive, and that the user is in possession of the necessary maintenance operations schedules and cards. The procedures described for the plant record and inventory section provide for the transfer of the record cards. Spares, other than minor items common to all locomotives, should be stocked and controlled at the Group's stores.

For a fleet of road vehicles, on the other hand, the quicker turn-round which is essential to maintain the standard of roadworthiness needed to comply with the regulations, and the fact that servicing should be carried out on a mileage basis for this type of equipment, entail a different approach to control. Two important principles should be followed; the first is that the transport section should operate and deploy the vehicles, and be responsible for the minor servicing normally carried out by the operational staff, such as greasing, cleaning, topping up, changing wheels, and keeping up tyre pressures. Further, they should report any defects noted. Secondly, the mechanical and electrical maintenance is essentially an engineering service to be controlled by the Group and the vehicles should be made available on demand, with the proviso that where it is essential to maintain the transport services, replacement vehicles are made available as necessary.

The control wall board illustrated in Fig. 5.20 can be built up from proprietary equipment, and the accompanying notes and Fig. 5.21 describe a method of control which satisfies the foregoing requirements.

Similar methods can be devised and established for other pool services of the Group. Consideration is next given to work involving the employment of resources external to the

GROUP POOL SERVICES - ROAD VEHICLES.
DEPLOYMENT & MAINTENANCE.

| INSPECT'N. REPORT FILE NO. | REGN. NO. | DESCRIPTION | MAKE | PAY LOAD | BODY | | PLANT NO. | | DEPLOYMENT | | | | | | | SERVICING DUE AT | | | LAST SERVICE DONE AT | NEXT SERVICE SCALE | | | | | MILEAGE | | |
|---|---|---|---|---|---|---|---|---|---|---|---|---|---|---|---|---|---|---|---|---|---|---|---|---|---|---|---|
| | | | | | PLAT-FORM | END TIPPER | OPS. CARD REF. | ITEM | A | B | C | D | E | F | G | 500 | 1000 | 3000 | | 500 | 1000 | 3000 | 6000 | 12000 | AT LAST SERVICE | NOW | DIFF. |

REPORTS FILED IN SEQUENCE NUMBERED FOLDERS IN ORDER OF DATES.

PLANT RECORDS PROCEDURE APPLIES.

NOTE:
ALL VEHICLES TO BE MAINTAINED TO A STANDARD OF ROAD-WORTHINESS COMPLYING WITH CURRENT REGULAT-IONS AND REQUIREMENTS OF THE APPROPRIATE AUTHORITIES.

| DUE IN AT | BREAK-DOWN. | REMARKS. |
|---|---|---|

| DEPLOYMENT KEY | | | |
|---|---|---|---|
| LOCATIONS. | | | |
| A | | E | |
| B | | F | SPARE |
| C | | G | CENTRAL SHOPS FOR SERVICING. |
| D | | | |

FIG. 5.20. Group pool services—road vehicles deployment and maintenance.

ROAD VEHICLES.
SERVICING ADVICE NOTE.

| FROM | GROUP. ENG^G. MAINTENANCE CONTROL | | TO | GROUP TRANSPORT SUPERINTENDENT. | |
|---|---|---|---|---|---|
| | ISSUED BY | DATE | | RECEIVED BY | DATE |
| | | | | | |

THE UNDERNOTED VEHICLE IS DUE AT GROUP WORKSHOPS FOR SERVICING ON THE DATE SHOWN :-

| LOCATION | DESCRIPTION | MAKE | CAPACITY. | REG^N. NO. | DUE AT WORKSHOPS |
|---|---|---|---|---|---|
| | | | | | |

THE FOLLOWING VEHICLE IS BEING ALLOCATED IN PLACE OF THE ABOVE :-

| DESCRIPTION | MAKE | CAPACITY | REG^N. NO. | AVAILABLE AT W'SHOPS ON | REMARKS. |
|---|---|---|---|---|---|
| | | | | | |

IF FOR ANY REASON, THIS CHANGE-OVER CANNOT BE EFFECTED, PLEASE NOTIFY THIS OFFICE WITHOUT DELAY, STATING WHY. THE NECESSARY "PLANT MOVEMENT AND RECEIPT ADVICE" COVERING BOTH DISPATCH AND RECEIPT MUST BE FORWARDED TO GROUP PLANT RECORDS OFFICE.

ROAD VEHICLES.
WEEKLY SUMMARY OF TOTAL MILEAGE READINGS.
AS AT FRIDAY.........DAY OF........19.....

| FROM | GROUP TRANSPORT SUPERINTENDENT. | | TO | GROUP ENG^G. MAINTENANCE CONTROL. | |
|---|---|---|---|---|---|
| | SIGNED | DATE | | RECEIVED BY. | DATE |
| | | | | | |

| VEHICLE REGISTRATION NO. | TOTAL MILEAGE READING | REMARKS FOR ATTENTION. |
|---|---|---|
| | | |
| | | |
| | | |
| | | |

IT IS ESSENTIAL THAT ANY FAULTS SUSPECTED OR NOTED IN THE WORKING OF THE RECORDING INSTRUMENT OR SPEEDOMETER ARE REPORTED TO GROUP ENG^G. MAINTENANCE CONTROL PROMPTLY FOR ATTENTION.

FIG. 5.21. Road vehicles maintenance schedule—associated documentation.

Group. As already noted, this should be controlled by the Group, to ensure that internal Group resources are fully used before external resources are employed.

It is often desirable and useful to arrange external maintenance contracts for specialised equipment, either with the makers, or with firms specialising in such work. Plant of this type includes instrumentation for compressed air plant, ventilation equipment, and boilers and power plant; testing machines and equipment; weighbridges; and passenger lifts. Such contracts usually take the form of an agreement whereby, for a nominal charge per visit, or per annum, an engineer inspects the equipment at agreed time intervals, and checks that it is in good working order. He carried out general servicing, making any adjustments or doing any minor repairs required, and reports in good time on repairs and replacements needed, or on matters which should be kept under observation. The cost of replacements and services on repairs are outside the contract fees, and are invoiced as and when they arise, after they have been agreed and the work carried out. Appendix III gives a typical maintenance contract of this type.

External resources (other than those involved in the supply of proprietary spares and renewals) which it may be necessary to use, can be summarised as follows:

Manufacturing parts for repairs to plant and equipment at site.

Repairs to plant, or processing of components or parts consigned from site.

Provision of personnel for site installation, repair, and overhaul work; these may be a supervising engineer only, skilled labour and supervision only, or total labour and supervision.

Provision of special tackle, equipment and tools for site work.

Rendering services under maintenance contracts for specialised equipment.

In all cases, the Group's engineering department should be made aware of the need, and satisfied that it is necessary to employ external resources. In some cases, the makers of the equipment, or their recognised servicing representatives, may be the only sources available. In other cases, it may be policy to call for competitive tenders for the materials and services required.

Orders should be drawn up by the Group's engineering staff in a clear and concise manner, and passed to the purchasing department for issue. The purchasing department should send copies of the order to the Group and Unit engineering offices for their information and reference. This also serves to confirm the placing of the order. It is useful for the Group's engineers to build up a reference file of external services available, and used from time to time, with the names of the firms, the types of service given, the persons to be approached, and the means of establishing day or night contact with them. This can be of particular value in an emergency.

## PRELIMINARY SCHEDULING AND ADJUSTMENT OF SCHEDULES; UNSCHEDULED WORK

At this stage it is possible to apportion to the Unit's maintenance schedules the work load arising from the work scheduled and controlled by the Group. This is done through co-ordination between the maintenance control centres of the Group and the Units, so that Group and Unit arrangements and activities are correlated.

To make up the Unit's master maintenance schedules, already referred to, the following procedure can be adopted. The data at the left of the base page of the Unit's activity maintenance schedule are first entered from the plant layout and flow diagrams already described. Next, the data in the first three columns of the schedule sheets are entered for each element of maintenance opposite each schedule on each base page. The first category of plant in the plant summary sheets is then considered, and using the maintenance pattern applicable, and allowing for Group control and holiday periods where appropriate, the patterns of the maintenance operations, their frequency and phasing, are entered on the appropriate schedules. All items in the schedules within this particular category, together with the related maintenance pattern, should be scheduled in this one operation. It is usually convenient to mount the maintenance pattern in a phasing slide (Fig. 5.6), with a double 52-week scale along the top edge. The maintenance pattern as a whole can then be set to any desired phasing and transferred to the appropriate schedules, while retaining the correct relation between the various operations. It is convenient to enter the phasing for the first item in the category under consideration, for all of its elements, and then to adjust the maintenance pattern by one or more weeks for each item in that category, till the latter is completely scheduled. Next, the man-hour loadings committed under each element of maintenance should be entered to scale on the element man-hour loading charts already described. This illustrates the distribution of the work load throughout the year.

The remaining types of plant in the plant summary sheets should then be similarly dealt with. To achieve a satisfactory distribution of work, the heaviest work-load of each maintenance pattern should be arranged to occur in the weeks with the lightest work load so far entered in the element man-hour loading charts; in this way, a reasonable degree of evening-out can be achieved. There will inevitably be irregularities in the weekly work load, but it will be possible to establish an average weekly level of scheduled man-hours which together with the allowances decided upon, will indicate the regular staffing necessary. Where the weekly man-hours exceed the average, the excess may be absorbed by the allowances provided for in the loading chart. Failing this, it may be met by the use of additional overtime for the zone labour force, the secondment of available labour from another zone of the Unit, the employment of the Group's resources, or the use of external facilities.

In preparing the element man-hour loading charts an effort should be made to level out the weeks above and below the average, by re-examining the jobs and tasks in those weeks, and re-scheduling them where possible to improve the load distribution.

## Unscheduled work

The loading charts just described include various allowances on capacity, in terms of manpower and time; together with a margin for unscheduled work. This gives some degree of latitude in the operation of the schedules generally, and absorbs unscheduled work in each Unit in the following categories.

### *Running repairs and adjustments arising from inspection*

An inspector's duty is to inspect, assess, and report, and the inspections are made to pre-set schedules. He should not be diverted, and his time should not be dissipated, in trying to effect running repairs and adjustments as he goes along. His primary tasks are to furnish, in good time, the information upon which action can be taken to provide the necessary attention, with a view to forestalling breakdown, and to provide guidance and supervision on the maintenance standards needed and achieved.

As the arrangements covering the scheduled elements of maintenance become effective (lubrication, routine servicing, site overhauls, and periodic replacements) the work load from this unscheduled element should diminish.

Defects noted from inspection should be classified in one of two ways. If the defects call for immediate attention, the inspector should contact the zone chargehand so that prompt action can be taken. Other defects can be discussed when the chargehand goes over the reports with the inspector; priorities can be allocated, and the items included in an additional weekly work schedule for this element of maintenance, or alternatively entered into the current schedules as extras at an opportune period. In doing this, reference should be made to the programmed work ahead; it may be that the plant in question is already scheduled for routine servicing, site overhaul or periodic replacement in the near future.

## *Breakdowns with group assistance on demand*

The term 'breakdown' is generally accepted as referring to a failure of plant or equipment which directly affects production and calls for immediate attention. The risk of breakdown is considerably reduced by thorough inspection, timely running repairs and adjustments, cleanliness, systematic and correct lubrication, regular servicing, and periodic overhauls or replacements. As the arrangements covering these elements of maintenance become more effective, the incidence of breakdowns diminishes, though they cannot be completely eliminated.

The effects of breakdowns can be limited by providing standby plant for essential services and operations, where justified economically.

The effects of breakdowns can be alleviated in several ways. They include:

Ready availability of spares and renewals, assemblies, or complete plant units. Thought should be given to the convenient locating of heavy and cumbersome parts, so that they can easily be moved into position in an emergency.

Making provision for easy access and means of handling.

Ensuring that rapid communication can be established at any time, with the transport and services of the Units, the Group, and the external suppliers.

Making the fullest use of labour employed on unscheduled work, or from the general services of the Unit or Group, before diverting labour from scheduled work in progress. On no account should labour be withdrawn from work in progress to meet statutory commitments.

When a breakdown does occur, the circumstances must be investigated and recorded, every effort must be made to establish the cause, and action should be taken where possible to obviate or lessen the chance of a recurrence. The demands of production may not permit full investigation and the effecting of modifications immediately, but the essential data and facts on which subsequent action can be based without interrupting production, should be established.

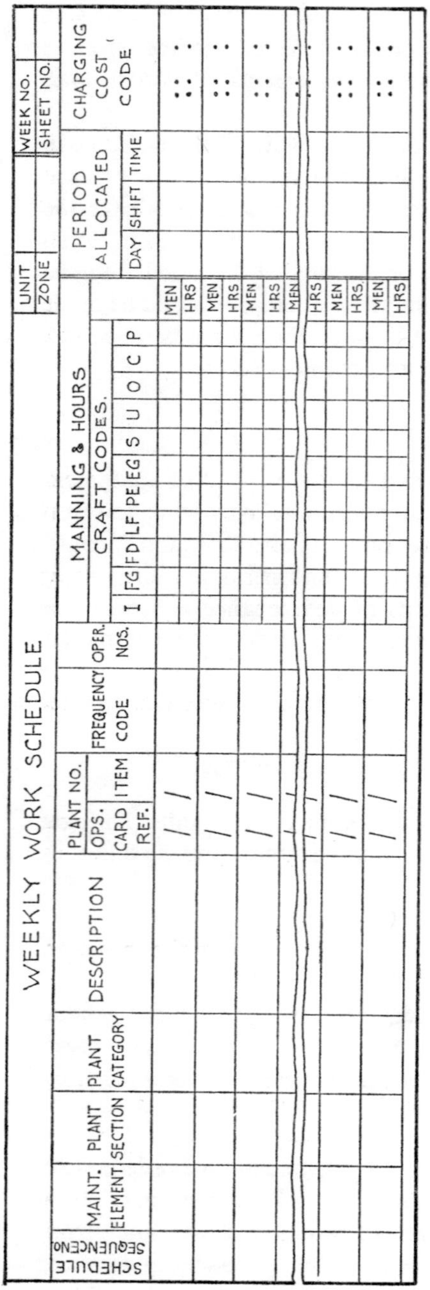


FIG. 5.22. Weekly work schedule.

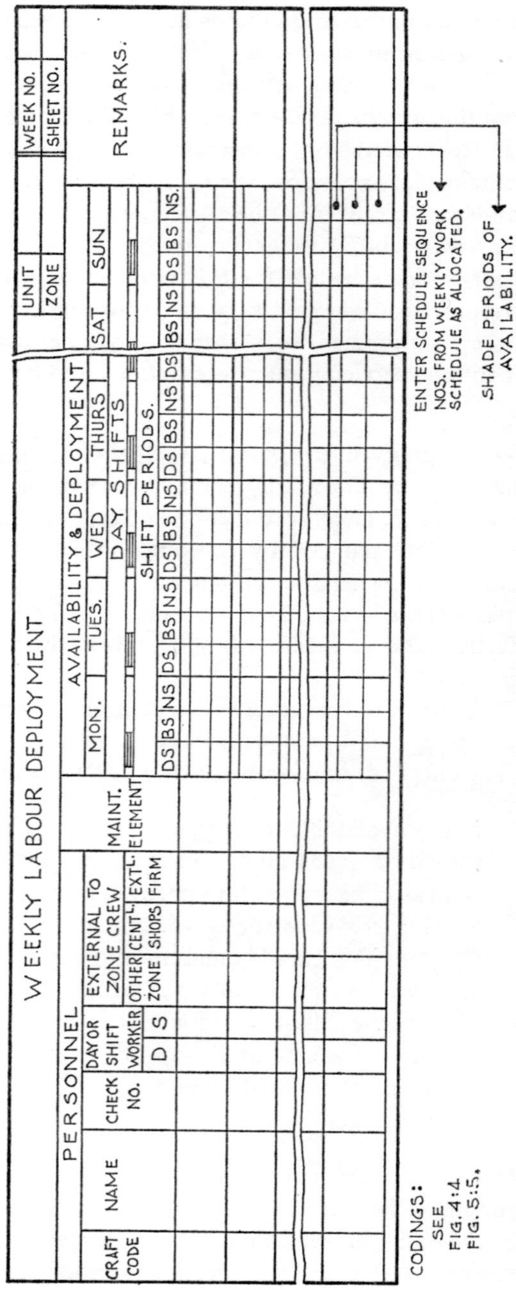


FIG. 5.23. Weekly labour deployment schedule.

## THE ALLOCATION OF TIMES

When the weekly maintenance load is known, it is necessary to allocate specific periods and time for the separate jobs and tasks in any particular week. Plant availability, in relation to standby conditions, scheduled running periods, and production operations, must be taken into account against the nature and time demands of the maintenance work. From the master maintenance schedules, the weekly work schedules can be prepared in the form illustrated in Fig. 5.22, suitably filed, and indexed under the Unit's identity and the maintenance zone covered.

## AVAILABILITY AND DEPLOYMENT OF PERSONNEL

This can be established by means of weekly labour deployment schedules of the form shown in Fig. 5.23. They can be drawn up in advance from the weekly work schedules, filed in a suitably indexed folder under the reference of the Unit and maintenance zone concerned.

## INSTRUCTING AND DIRECTING PERSONNEL

This is effected by setting up, side by side, the weekly work schedule and weekly labour deployment schedule, and issuing to the operative, the maintenance operations cards (Fig. 5.24) appropriate to his daily work. The latter should be indexed and filed for easy reference and selection, and their issue and return controlled by the maintenance control section. The latter should display the weekly work schedule (Fig. 5.22) and the weekly labour deployment schedule (Fig. 5.23) towards the end of the week preceding that to which they apply, and remove them at the beginning of the week following that to which they apply.

### Unscheduled work

All unscheduled work and demands made upon the site engineering establishment, as distinct from the work covered in the planned schedules, should have some form of simple documentation equivalent to the master works order for the central workshops (Fig. 4.12). This can take the form of the 'service request form' shown in Fig. 5.25. This is the initiating and authorising document for all such demands, and is issued by all sections within the Unit requiring the services of the Unit's engineering department. The pre-printed layout provides for all the needed data, including charging instructions and costs. Authorised users are issued with books of say 50 duplicate 'service requests'. The pages are numbered in duplicate, with a tear-off top copy; the carbon copy is retained in the book, which can be indexed and filed for reference.

All service requests should be sent to the Unit's engineering office for authorisation and action. Enquiries regarding progress should be made through the same office, quoting the originating source and the service request number from the carbon. Sketches should be made on separate sheets from sketch books of the type described in Chapter 4. All necessary information should be given on the 'service request'. If the drawing number or pattern

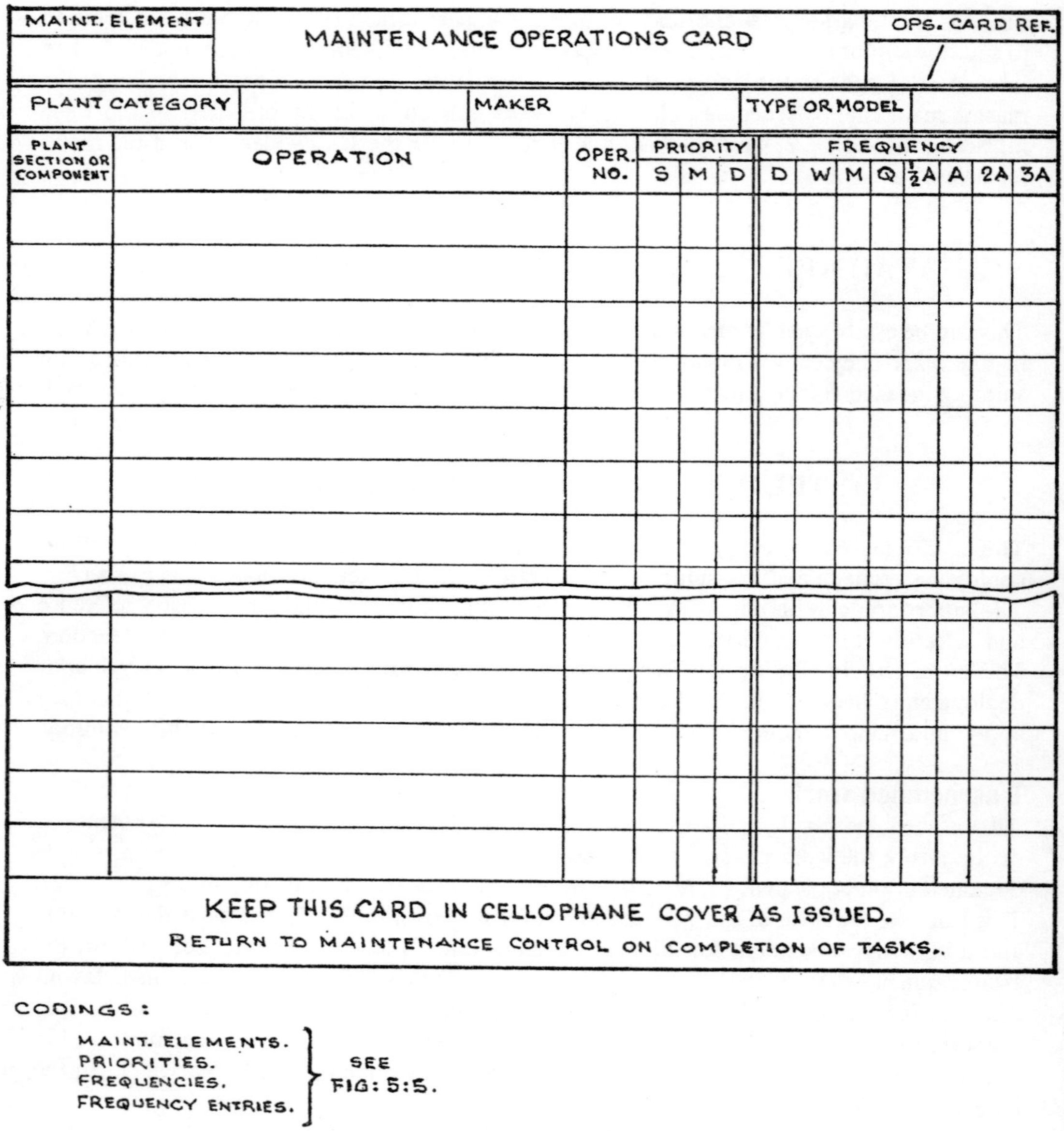

| MAINT. ELEMENT | MAINTENANCE OPERATIONS CARD | OPS. CARD REF. |
|---|---|---|
| | | / |

| PLANT CATEGORY | | MAKER | TYPE OR MODEL | |
|---|---|---|---|---|

| PLANT SECTION OR COMPONENT | OPERATION | OPER. NO. | PRIORITY | | | FREQUENCY | | | | | | | |
|---|---|---|---|---|---|---|---|---|---|---|---|---|---|
| | | | S | M | D | D | W | M | Q | ½A | A | 2A | 3A |

KEEP THIS CARD IN CELLOPHANE COVER AS ISSUED.
RETURN TO MAINTENANCE CONTROL ON COMPLETION OF TASKS.

CODINGS:
MAINT. ELEMENTS.
PRIORITIES.
FREQUENCIES.
FREQUENCY ENTRIES.
SEE FIG: 5:5.

FIG. 5.24. Maintenance operations card.

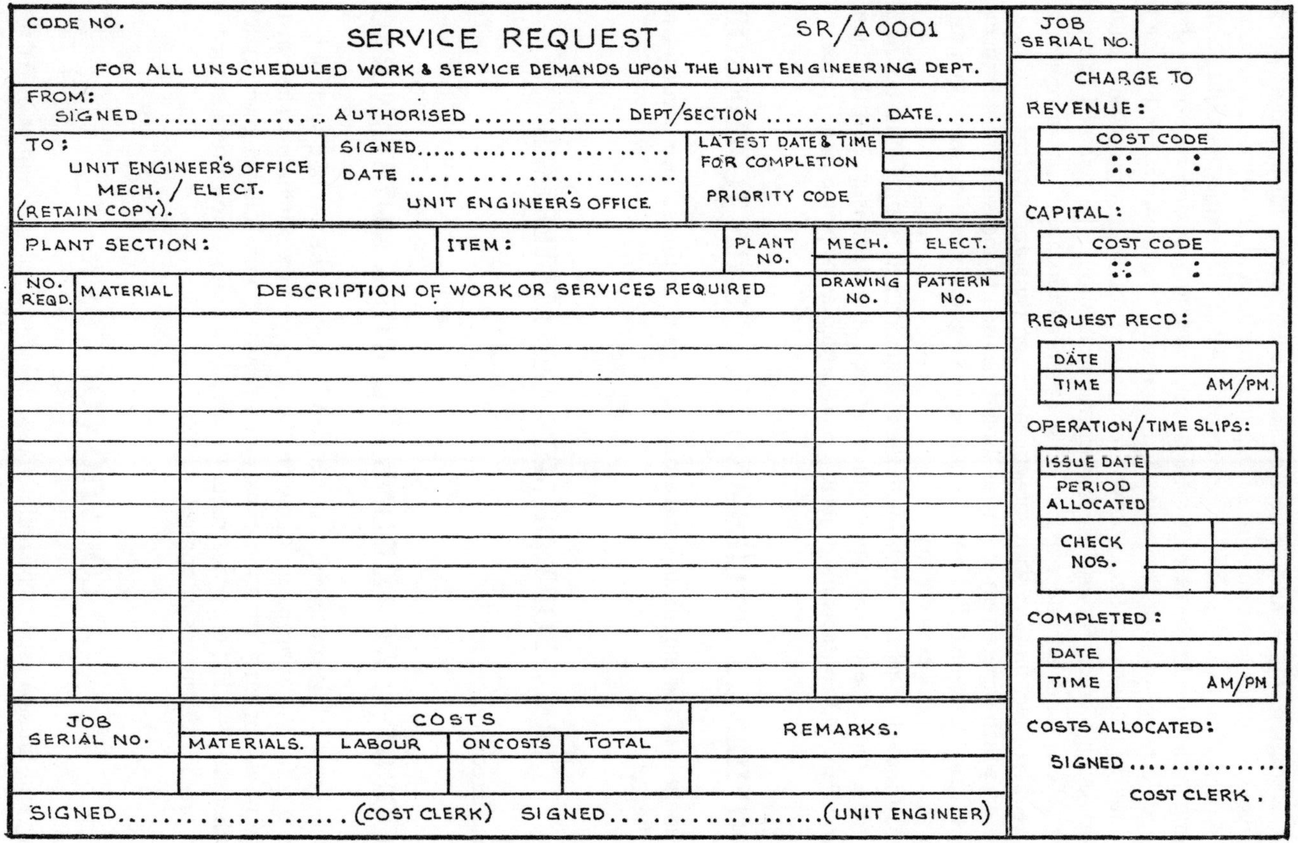

CODE NO. SERVICE REQUEST SR/A0001

FOR ALL UNSCHEDULED WORK & SERVICE DEMANDS UPON THE UNIT ENGINEERING DEPT.

FROM:
SIGNED ................ AUTHORISED ............. DEPT/SECTION .......... DATE......

TO: UNIT ENGINEER'S OFFICE MECH. / ELECT. (RETAIN COPY).

SIGNED......................... DATE ......................... UNIT ENGINEER'S OFFICE

LATEST DATE & TIME FOR COMPLETION

PRIORITY CODE

PLANT SECTION: ITEM: PLANT NO. MECH. ELECT.

| NO. REQD. | MATERIAL | DESCRIPTION OF WORK OR SERVICES REQUIRED | DRAWING NO. | PATTERN NO. |
|---|---|---|---|---|
| | | | | |

| JOB SERIAL NO. | COSTS | | | | REMARKS. |
|---|---|---|---|---|---|
| | MATERIALS. | LABOUR | ONCOSTS | TOTAL | |
| | | | | | |

SIGNED.................. (COST CLERK) SIGNED.................(UNIT ENGINEER)

JOB SERIAL NO.

CHARGE TO

REVENUE:

COST CODE

CAPITAL:

COST CODE

REQUEST RECD:

DATE

TIME AM/PM.

OPERATION/TIME SLIPS:

ISSUE DATE

PERIOD ALLOCATED

CHECK NOS.

COMPLETED:

DATE

TIME AM/PM

COSTS ALLOCATED:

SIGNED ..............

COST CLERK.

FIG. 5.25. Service request form (unscheduled work).

number is known, it should be entered. The latest date for completion when applicable, is IMPORTANT, and should be genuinely assessed, to assist in the correct allocation of priorities to demands. Requirements should be determined and advised as far ahead as practicable, to reduce urgent demands to a minimum. The plant number should be given where possible, to establish correlation with the Unit's plant records. All service requests should be initialled by the issuing authority. In the case of breakdown, the Unit's engineering office should be contacted and details given to enable action to be initiated; the service request which follows should be marked 'URGENT BREAKDOWN—CONFIRMATION'. This procedure should be reserved for genuine emergencies. The service requests should be retained in the Unit's engineering office, and filed under a job serial number (allocated from a register) which relates the request sheet number with subsequent progressing and control.

The scale on which service requests arise should be used to measure the adequacy of the scheduling already described. The aim should be to keep unscheduled work to a minimum by revising the schedules as necessary.

The issue of these forward instructions to the maintenance personnel must not be allowed to detract from the close day-by-day contact between the supervisors and the men under their control. Personal contacts must always be regarded as of prime importance.

## PROVIDING SUPPLIES AND MATERIALS TO THE JOB

This is a function of the purchasing and stores department; this service is considered as a whole in Chapter 6.

## TRANSPORT OF MEN AND MATERIALS

This is considered in Chapter 7.

## RECORDING WORK DONE AND WORK OUTSTANDING

Where scheduled work is concerned this can be achieved by the use of a symbol indicating completion against the appropriate items on the weekly work schedules. The outstanding items can be noted for early attention or carried forward into the following weekly schedule.

For unscheduled work, the service requests already described constitute in themselves a record and provide the means of control.

## ALLOCATION OF COSTS

The costing and finance aspects of maintenance are discussed in Chapter 8. At the Unit, the medium of labour cost allocation to the job can be similar to that at the central workshops, using the clock card (Fig. 4.25) and the time allocation card (Fig. 4.24). The allocation of material charges to the job is effected by means of the stores warrant (Fig. 6.12) and the stores credit note (Fig. 6.13). These and any other charges arising at Unit level are linked up with the general costing procedures and policies set out in Chapter 8.

## ASSESSING THE OPERATIONAL FACILITIES AND SERVICES REQUIRED

These are determined as the processes of organisation at Unit level are established step-by-step; they become evident as this work progresses. The consideration given to the central workshops in Chapter 4 serves also as a guide to the assessment of the Unit's facilities. The provisions are on a smaller scale, of course, and take into account the facilities at the central workshops, full advantage of which should be taken. These facilities and services may be augmented through the medium of the works order (Fig. 4.12) directed through the Group's engineering office, as described in Chapter 4. This ensures that the Group's resources are used to full advantage, before engaging external services. A feature that may deserve consideration in this connection is the provision of a mobile workshop in the form of a van equipped to aid the site services on specialised work.

A typical schedule of a Unit's facilities might include:

*Machine tools*

Small gap bed centre lathe, radial drill, pillar drill with chuck attachment, pipe screwing machine, tube bender, machine saw, tool grinder.

*Plate shop*

Portable electric welder, gas welding and burning equipment, straightening machine.

*Smithy*

Down-draught hearth, anvil and swage block.

*General*

Degreasing equipment, work benches, marking-off table, compressed air service, tools and tackle.

## ACCOMMODATION REQUIRED AND LAY-OUT

The maintenance accommodation required at the Unit, and its lay-out, will be governed by the organisational structure of the undertaking concerned. As mentioned in Chapter 1, a composite or Group industrial undertaking is selected as the basis for consideration here; this provides a central engineering organisation at Group level, with extensions to suit the operation and control of the engineering services at the separate plants or Units. Such extensions would include, either as separate set-ups under the Unit's engineer, or as a unified set-up at his location, the engineering office, the maintenance control centre, the plant record

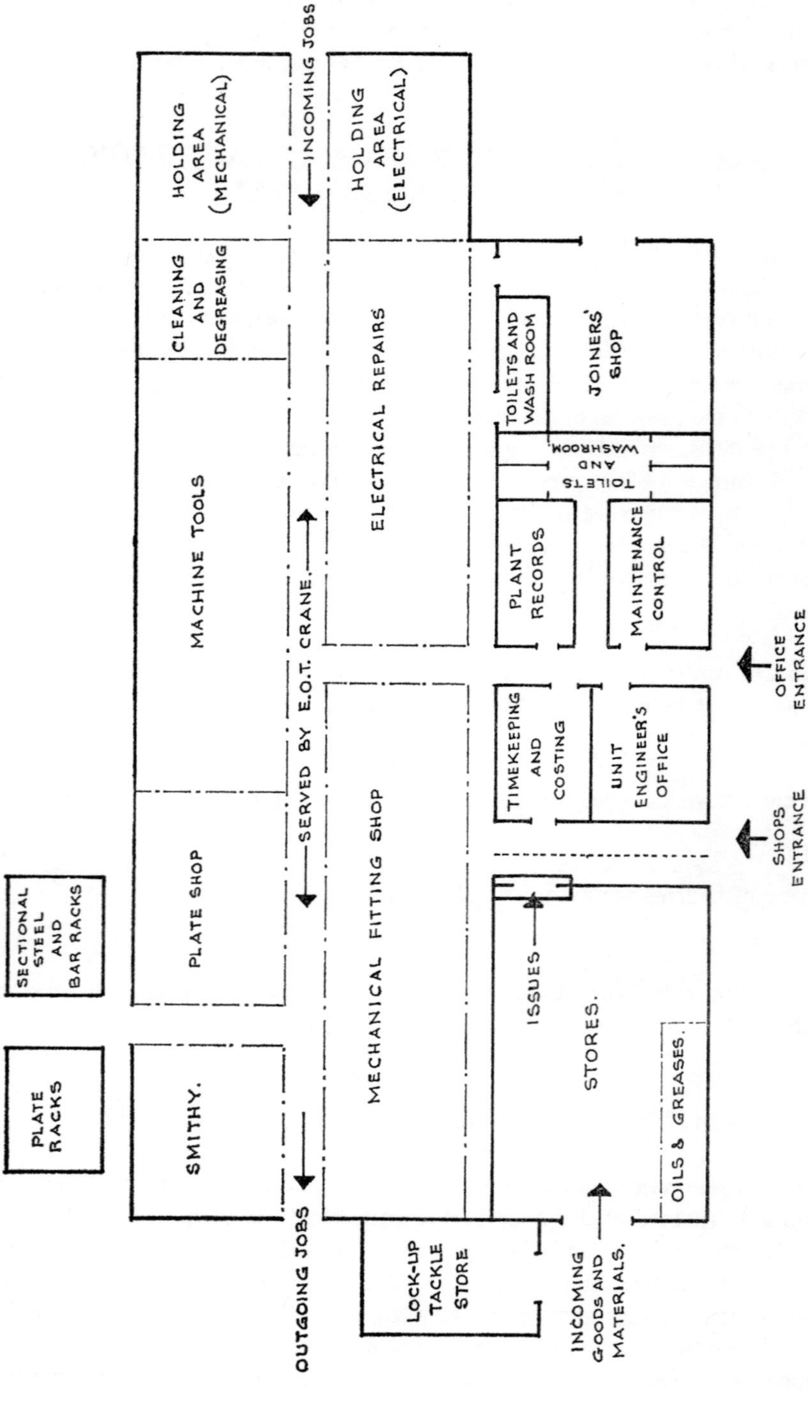


FIG. 5.26. Typical layout of maintenance accommodation at a unit activity.

section, and the Unit's workshop. In any event, the control is improved if these are accommodated under one roof. A suggested layout of accommodation, allied to stores accommodation and cost office, is illustrated in Fig. 5.26.

## DOCUMENTATION AND OPERATIONAL PROCEDURE AND CONTROL

The whole of the administrative procedures and practices considered in the foregoing converge at each Unit. They form the engineering administrative environment which is designed to serve production, and production is centred at the Units. Hence it is appropriate to prepare a summary of the documentation and procedures, under the various elements of administration, with convenient references, related to the day-to-day operation of the Unit. The total maintenance arrangements for the Unit and the Group consist of the sum of the zone programmes and procedures involved in each case.

*Chapter 6*

# PURCHASING, STOREKEEPING AND WAREHOUSING
## for Industrial Plant Maintenance

The function of the purchasing and stores department is to supply equipment, materials, and supplies to the work. Materials are of major importance to the maintenance process, and account for a large part of maintenance costs. In large and small undertakings alike, the control of materials and stores is important to the economic success of maintenance. The organisation and procedures adopted in purchasing and stores departments vary considerably; in a large undertaking they become highly developed, and involve many considerations beyond the province of this book. However, certain fundamental procedures must be established, in order that the procedures for maintenance may be satisfactorily linked with this supporting service.

The typical organisation considered here is illustrated in Fig. 1.4 and referred to in Fig. 5.1. Purchasing and stores management is there shown as a Group function, working through branch stores at the central maintenance workshops and the various Units.

Maintenance engineers require the purchasing and stores department to provide and operate a number of services; these include:

Provision from stock of the necessary stores and sundries for a range of items as required.

Provision from stock of the plant items, spare parts, and components needed to service and maintain the plant and equipment.

Procurement of plant and materials required in the maintenance processes which are not kept in stock.

Employment of external assistance when required to supplement internal resources.

Provision of suitable storage and safe custody for spare plant and materials holdings, with easy location and identification, within the confines of the stores themselves or at selected sites within the premises. It is also necessary to ensure that the plant and materials are kept in satisfactory condition during storage.

Establishment (in collaboration with the maintenance engineering department) of means for controlling the ordering, checking, stocking, issuing, movement, transfer, disposal, crediting and costing of plant, materials, and services.

Delivery of plant, materials, and supplies between the external sources and either the Group's or Unit's stores, or directly to the point of usage; or between the Group's and

Unit's stores; or between the Group's or Unit's stores and the point of usage (except where the user department collects by hand).

Arrangement for access to the stores and provision of essential materials outside normal working hours, in cases of breakdown or emergency.

Collection, disposal, and crediting of all redundant and obsolete plant and scrap materials.

Collaboration with the maintenance engineering department in ensuring that the procedures and practices are compatible with those set up to carry out the maintenance engineering functions.

The purchasing and stores department requires the maintenance engineers to assist these operations in several ways. These include:

Collaboration to establish suitable holdings of essential stores and sundry items, plant items, spares and components, under suitable classifications, and with means of identification, and to indicate the expected rate of use so as to determine maximum and minimum stock quantities and forward ordering requirements appropriate to the delivery periods involved.

Provision of engineering data needed to procure the required plant, materials, and services.

Provision of a written list of signatories and authorities within the engineering organisation, including classification of stores covered by the authorisation, for demands made upon the purchasing and stores department.

Provision of the services needed to examine or inspect incoming equipment—either at the makers' works, before dispatch, or on receipt, or both—for condition, quality, workmanship and compliance with specification (where applicable).

Advice on the location and storage of engineering plant and equipment, and its preservation during storage.

Timely notification of requirements.

Assistance in correlating the procedures and practices of the purchasing and stores department with those of the maintenance engineering organisation.

Provision to the purchasing and stores department of those services which are operated by the plant record and inventory section.

Collaboration as necessary during stocktaking.

Notification of plant available for disposal as redundant, obsolete, or scrap; segregation and collection of scrap metals for transport and disposal.

## THE GENERAL ORGANISATIONAL FRAMEWORK

Figure 6.1 illustrates a typical framework in the type of industrial organisation considered in this book. The functional correlation with the engineering services and other sections involved, is developed later in this chapter.

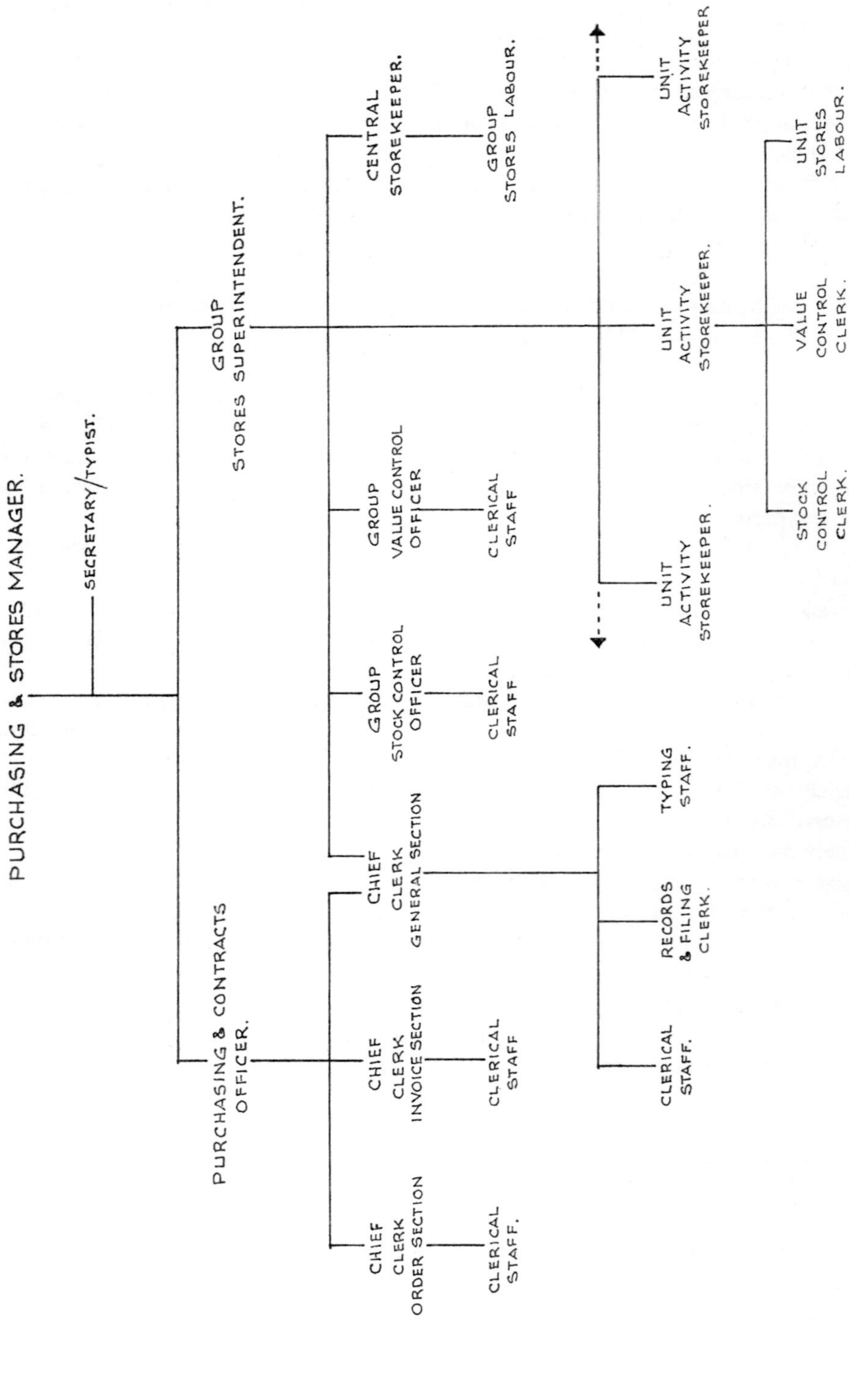


FIG. 6.1. Purchasing and stores service—general organisational framework.

## THE PURCHASING AND CONTRACTS SECTION

Purchasing and stores management is a Group function. Functionally, purchasing (as distinct from stores and storekeeping) is confined to the Group's organisation. The department comprises three main sections, namely:

### (1) The order section

This arranges the procurement from external sources of all plant, equipment, and materials, from the time the demand arises. It receives the users' demands, issues enquiries, receives tenders, and collaborates with the users' authorities in the examination and judgement of offers received. The section ensures that the approval of the appropriate technical authority is given, and issues and distributes copies of the official order or contract documents, incorporating the terms and conditions applicable. The progress of orders or contracts is followed, and delivery or completion urged as necessary. The disposal of all scrap, including obsolete and redundant plant, is arranged through the section; and it maintains liaison with the technical departments, and initiates direct technical discussions with suppliers or contractors when necessary. It also maintains liaison with the finance department.

### (2) The invoice section

This section receives all invoices and accounts from external sources for goods and materials supplied, and for services rendered. It is responsible for checking the accuracy and validity of invoices and accounts against the official order or contract documents, examining the stores advice notes to check the quantity and condition of the goods and materials involved, and involving the users' technical authorities in the inspection of materials and workmanship, securing the necessary approval signatures. It maintains liaison with the finance department.

### (3) The general section

This section provides such general services in the purchasing department as typing, correspondence, filing, and records. It also gives a common service in matters which relate to the activities of the purchasing and stores department as a whole.

## STAFF DUTIES AND RESPONSIBILITIES

Figure 6.1 shows a typical staff chart for the foregoing aspects of purchasing, and for the related stores organisation described later.

### The purchasing and stores manager

The responsibility for providing equipment, materials, and supplies to the job lies with the purchasing and stores manager, who is located with his establishment, within the Group's organisation. He is responsible to the Group's management for the control and supervision of the purchasing and stores department. A substantial part of the expenditure of the undertaking is routed through this section of its organisation. A short-sighted purchasing policy or faulty administration or slackness in this department can create large consequential losses, difficulties, and inefficiencies in many sections of the undertaking. It follows that this is an important appointment, and its occupant must be a man with a keen sense of 'buyer-craft', appreciating especially that material with the lowest first cost is not necessarily the cheapest in the end. He must remember, in the words of Ruskin, that 'There is hardly anything in the world that some man cannot make a little worse, and sell a little cheaper, and the people who consider price alone are this man's lawful prey'.

He must also appreciate that purchasing is not an isolated function, but is influenced by a number of indirect factors which involve close and constant collaboration with the technical and other departments of the Group. Although he is not a production official, it is part of his duty to interpret the needs of production in terms of equipment, materials, and supplies, and to ensure that prices and costs are maintained at a truly economic level. On the other hand, he must guard against any tendency for cheapness to develop into a fetish.

### The purchasing and contracts officer

The purchasing and contracts officer is responsible to the purchasing and stores manager for the work of the purchasing and contracts section of the department, comprising the order section, the invoice section, and the general section, with the proviso that the general section gives a common service to the whole purchasing and stores department.

### Section chief clerks

These are responsible to the purchasing and contracts officer for the allocation, supervision, and co-ordination of the work in their respective sections, and for maintaining the necessary records.

## PURCHASING DOCUMENTATION

Forms and stationery employed in purchasing procedures and practices naturally vary widely. The essential feature should be that each document fulfils a necessary purpose or provides information not otherwise available; this is not always the case in practice. The key-notes should be simplicity and convenience in lay-out; the use of colours to facilitate rapid recognition; a clear understanding of the routing; speedy handling; and systematic recording and filing.

The following discussion is restricted to documents which are required for the maintenance aspect of purchasing, and typical designs which fulfil the joint purposes of the departments concerned are considered. The illustrations are confined to documents exchanged internally between the purchasing and stores department and the production/engineering and finance departments, and exclude those involved in transactions between the purchasing department and external suppliers or contractors. They are:

| | |
|---|---|
| Internal Request to obtain Tenders | — Fig. 6.4 |
| Tender Summary | — Fig. 6.5 |
| Internal Revenue Requisition | — Fig. 6.6 |
| Internal Capital Requisition | — Fig. 6.7 |
| Application for Plant, Equipment or Services under Capital Expenditure | — Fig. 6.8 |
| Internal Urging Note | — Fig. 6.9 |

# STORES AND STOREKEEPING

## (1) General stores and sundries

These comprise general stock items and miscellaneous goods in common and regular use, usually supplied from the Group's stores to the various Units' stores, to replenish their stocks at regular intervals. The stock holdings at the Units are related to the rate of usage, and should be kept as low as reasonably possible. The range of general stock items and materials is too wide to list here, and the nature of the industry served also affects the nature and quantity of the stocks. Some examples are given in the next section.

## (2) Plant sub-assemblies, spare parts and components

These comprise parts which are to be built into existing plant items and equipment, in the course of the maintenance and servicing processes. They must be clearly identified with the parent apparatus of which they are an integral part, and this entails accurate correlation with the actual plant or machine, the manufacturer, and the latter's machine numbers and part numbers; allowance must also be made for interchangeability of parts between similar machines.

The range of such items is usually very wide, and their physical proportions in terms of size and weight can also vary widely. In principle, items common to more than one Unit should be stored centrally, and items peculiar to a particular Unit should be stored at that Unit. Those items which can be stored in open enclosed compounds, with protective treatment where necessary, should be differentiated from items which should be housed within the precincts of the stores proper. Particularly cumbersome sub-assemblies and spares, with their handling facilities, should be located as near as possible to their point of use, on the works premises, so that in the event of emergency or breakdown, they can be moved readily into position, with the minimum loss of production.

The compiling of the inventory for this type of stores is helped by (and based on) the information and data available from the plant record and inventory section (see Chapter 2).

## (3) Plant units or composite assemblies

These are capital equipment, or held in a suspense account and capitalised on putting to use. Notwithstanding that they may be built into composite installations, they are complete units of equipment in themselves, such as a pump/motor/bedplate assembly; a compressor; a conveyor; an electric motor; or a transformer.

Though the range and number of items involved is more limited, there is still a wide diversity in their sizes and weights, and they are usually larger than plant sub-assemblies, spare parts, and components. As with the latter category, items used at more than one Unit should be stored centrally, while those items peculiar to a particular Unit should be stored at that Unit; and items which can be stored in open enclosed compounds, with protective treatment where necessary, should be differentiated from those which should be located on the works premises, and those to be housed within the stores.

In compiling the capitalised or suspense account holding in this category, recourse to the plant record and inventory section described in Chapter 2 is essential. Unlike the other types of stores described, this class is controlled by the plant record and inventory section, and the functions of the latter which bear upon purchasing, stores, and finance, can apply to equipment of this type.

## ESTABLISHMENT OF STORES INVENTORIES: CLASSIFICATIONS, CODINGS, PARTS LISTS AND HOLDINGS

The primary classification of stores holdings can be based on the main categories already set out, namely:

(1) General Stores and Sundries.
(2) Plant Sub-assemblies, Spare Parts, and Components.
(3) Plant Units or Composite Assemblies.

These can then be subdivided in a manner suited to the nature oт the materials, and the requirements of operation and control.

### (1) General stores and sundries

The wide range of holdings under this heading makes it convenient to classify these in alphabetical order of description. The following is a typical breakdown:

1/01 — Building Requisites.
1/02 — Castings (Ferrous).
1/03 — Castings (Non-ferrous).
1/04 — Electrical Supplies and Accessories (as distinct from plant).
1/05 — Oils and Greases.
1/06 — Pipes and Pipe Fittings.
1/07 — Steel Sections, Bars, and Plates.
1/08 — Timber.
1/09 — Wire Ropes.
1/10 — Miscellaneous Stores and Sundries.

These sub-classifications can in turn be divided again to identify the articles or materials in more detail:

1/07/10 — Angles, Rolled Mild Steel.
1/07/20 — Channels, Rolled Mild Steel.
1/07/30 — Flats, Mild Steel.
1/10/10 — Belting.
1/10/20 — Bolts and Nuts.
1/10/30 — Brooms and Brushes.

To these main and sub-classifications can then be added a sequence number, so that the total coding is established for the complete identification of an item:

1/07/30/001 — 50 mm × 10 mm MS Flat.
1/07/30/002 — 50 mm × 15 mm MS Flat.
1/07/30/003 — 50 mm × 20 mm MS Flat.
1/10/20/004 — 12 mm × 75 mm Bolts and Nuts, MS Black, HRH.

An indexed 'vocabulary' of general stores and sundries can be compiled on this basis, in loose-leaf form, for use at all stock-holding points making demands on the stores for this type of stock. Such coding ensures that similar items or materials are identified in like manner at every stock-holding point, and helps to produce clearly specified requirements, orderly storing and locating, and good stock control. A typical page (flat A4 for preference) is shown in Fig. 6.2 (part).

### (2) Plant sub-assemblies, spare parts and components

For stocks of this type, the plant record and inventory described in Chapter 2 forms a good framework on which the inventory of spare parts and components can be built up. This latter process is largely a logical development of the plant record and inventory, which is concerned with plant items and equipment; it entails a further division (into component parts) to establish the inventory for this class of stock.

The first sub-division can be related to the types of plant and equipment which form the basis of the plant record and inventory coding, working through these in alphabetical order:

2/01 — Air Receivers.
2/02 — Boilers.
2/03 — Concrete Mixers.
2/04 — Compressors.
2/05 — Conveyors.

Within each of these divisions, there may be plant from different makers, with differing component parts. It is also reasonable to assume that any interchangeability of parts between the items in any one division will be confined to items from the same manufacturer; even though the detailed design may be different, or may have been changed.

With this in mind, the next sub-division can usefully be based on the makers' names, in alphabetical order:

2/04/01 — Compressor (ABC & Co. Ltd).
2/04/02 — Compressor (DEF & Co. Ltd).

Since plant of any one type, bought from one maker, may still include a series of models or types, this factor must be allowed for in the final codings and listings. The incidence of interchangeability between these models or types must also be included, so that maximum coverage is provided for the minimum cost. Hence, to complete the coding, the component parts are identified by adding numbers in sequence throughout the range of a maker's products in a particular plant category. The final coding is related to the maker's part number, and includes the model or type in which the part is used. For parts common to more than one model or type, the common coding chosen should be shown separately, so that identical parts and components can be identified, stored, and withdrawn against the same stores coding. Where a common coding is referred to, the normal coding space in the records should be left blank. The common coding can conveniently be the coding employed for the component the first time it occurs in the list, this being referred to (in a common coding column) each time that the component recurs, as the records are compiled.

This process involves much work and close collaboration with the makers, to identify clearly all the components used in their products, with standard descriptions and makers' references, and allowance for design changes which may affect the holdings for machines manufactured at different periods. Full use should be made of makers' illustrations and parts lists, which should be filed separately, indexed, and correlated with the lists.

Figure 6.2 (part) illustrates the form and build up of a typical list for this type of stores, taking the foregoing features into account. It is preferably kept in loose-leaf form, to simplify the up-dating of the records. Abridged versions, appropriate to the plant in use and to stores held locally, should be available at the various Units. Alternatively, spares catalogues can be made up from the list, including only the spares holdings, as distinct from every part and component in the list. In any event, distribution should be made only to essential points, and a clear-cut procedure and responsibility should be established for keeping the records up-to-date and maintaining liaison between all sections of the organisation, and the manufacturers.

## Explanatory notes for page 2 of list (see Fig. 6.2)

### *The group list (stores type* 2)

The Group list is developed from the plant record and inventory, and is an inventory of all the component parts of the plant and equipment included in the latter. It differs from the spares catalogue in that it identifies each part and component, as distinct from spares holdings only. From it are derived the spares catalogues appropriate to the various stocking and user points, allowing for developments in design, interchangeability, and necessary stock levels.

### *The index page*

This gives a summary of the plant or machine, for a given type and manufacturer. It records the various sizes and models, with their year of manufacture (to assist in establishing interchangeability). References to the makers' illustrated parts lists are included, as are the makers' numbers and internal plant numbers appropriate to each model or type and the period of manufacture. It also gives the page numbers relating to component parts for the various models or types listed.

### *The parts lists*

These follow the index page, in the order given thereon. Even though some parts may be common to two or more of the machines covered, a complete parts list should be inserted for each model and period reference, so that the stores codings and user demands can be readily related to a change in design of any model or component part, and reference made to a common stores coding. A dash should be entered in the stores coding column when the part is already coded under a previous model or machine, and the earlier coding should be inserted in the common coding column. This also applies to STM/4 parts referred to below; in this case the common coding is that from the general stores and sundries schedules.

### *Maintenance class*

This section relates to stock level policy and also distinguishes those components which are classed as general stores and sundries. The two designations are STM (for Short Term

CLASSIFICATION 1 - GENERAL STORES & SUNDRIES:-

| FLAT MILD STEEL | | | |
|---|---|---|---|
| WIDTH | THICKNESS | CODING | |
| 2" | 1/4" | 1/07/30/001 | |
| | 3/8" | /002 | |
| | 1/2" | /003 | 07 |

PAGES ARRANGED IN CODING SEQUENCE.

CLASSIFICATION 2 - MACHINERY SUB-ASSEMBLIES, PARTS & COMPONENTS:-

TITLE PAGE:

| | |
|---|---|
| BINDER NO. | 1 |
| STORES MAIN CLASSIFICATION | 2/ |
| CATEGORY SUB-CLASSIFICATION | /01-05 |

PAGE 2:

EXPLANATORY NOTES AS SET OUT IN TEXT OF THIS CHAPTER.

INDEX PAGE:

| INDEX PAGE | | 2/04/01 |
|---|---|---|
| MAIN CLASSIFICATION | 2/ | SUB-ASSEMBLIES, PARTS & COMPONENTS. |
| SUB-CLASSIFICATION. | 04/ | COMPRESSORS. |
| MANUFACTURER'S CODING | 01 | ABC & CO. LTD |

| RANGE OF MAKER'S MACHINES USED | | | PARTS LISTS. | | MAKER'S NUMBERS RANGE | INTERNAL PLANT NO. RANGE | REMARKS |
|---|---|---|---|---|---|---|---|
| TYPE OR SERIES | PERIOD OF MAKE | | MAKER'S REF | INTERNAL INDEX NO. | | | |
| | YEAR | INDEX | | | | | |
| | | | | | | | |
| | | | | | | | |

PARTS LIST PAGE:

| | | |
|---|---|---|
| MACHINE OR PLANT | COMPRESSOR | PARTS LIST. |
| MAKER | ABC. & CO. LTD. | |
| MODEL OR TYPE | XYZ. | |
| MOBILITY CLASS. | MOBILE | |

| DESCRIPTION OF PART. | NUMBER | | UNIT OF ISSUE | | MAKERS' PART No. | MAKERS' DRG. No. | COMMON TO | | STORES CODING | COMMON CODING | MAINTENANCE CLASS | | | | |
|---|---|---|---|---|---|---|---|---|---|---|---|---|---|---|---|
| | PER SET. | PER M/C. | SINGLE | SUB-ASSEMBLY (BRACKETED) | | | MODEL | PERIOD | | | STM | | | | LTM |
| | | | | | | | | | | | 1 | 2 | 3 | 4. | |
| | | | | | | | | | | | | | | | |
| | | | | | | | | | | | | | | | |

| | |
|---|---|
| STORES MAIN CLASSIFICATION | 2 |
| CATEGORY SUB-CLASSIFICATION | 04 |
| MANUFACTURER'S CODING | 01 |
| MODEL & PERIOD INDEX | A |
| SHEET NO. | 1 |

| ESTIMATED LIFE | | STOCK LEVEL DATA | | | | STORING LOCATION | | | | | REMARKS |
|---|---|---|---|---|---|---|---|---|---|---|---|
| YRS. | MTHS. | USAGE RATE | DELIVERY (MTHS) | MAX. STOCK. | MIN. STOCK. | ACTIVITY | SECTION | BIN | AREA | COMPOUND | |
| | | | | | | | | | | | |
| | | | | | | | | | | | |

FIG. 6.2. Make-up of stores lists—classifications 1 and 2.

Maintenance parts) and LTM (for Long Term Maintenance parts). Short-term maintenance parts are further divided into STM/1 (life up to 6 months); STM/2 (life of 6 to 24 months); STM/3 (life of 24 months to say 5 to 10 years); and STM/4 (sundry parts, such as pins, bolts and nuts, keys, motor brushes, jointing, belting). Long-term maintenance parts are those which fall outside the foregoing categories; for example, non-wearing parts of machinery such as housings, frames, casings, and bedplates.

### *Estimated life*

This can be useful in certain cases.

### *Stock levels data*

This section requires close collaboration between the maker, the purchasing and stores department, and the user department within the undertaking. Reference should be made to the section on Build-up and Forward Availability of Spares and Renewals in Chapter 4.

### *Storing location*

This identifies the place of storage and the bin or compound location where applicable. It also serves to clarify delivery instructions to manufacturers and suppliers.

## (3) Plant units or composite assemblies

As indicated previously, these are essentially items of capital equipment; they should not be confused with the workshops pool of plant referred to in Chapter 4, which is essentially plant in circulation. The plant and equipment will usually come under the following headings, which should form part of the classifications and codings established:

| Classification | Stock Holding Code |
|---|---|
| Emergency stock items—loaned out and returned or replaced on issue | 01 |
| Plant held for projects and developments | 02 |
| Items held to cover short-notice additional demands of production | 03 |
| Replacement items held against the writing-off of plant in operation | 04 |
| Redundant plant for transfer or disposal | 05 |

This category is controlled by the plant record and inventory section, and the list must be prepared in such a way that it presents a continuous record of stock holdings, issues, and replacements, correlated with the plant record for separate items. This ensures that the identity of the plant is recorded through the allocation and attachment of plant numbers on receipt, and through subsequent verifications of current stock holdings and valuations.

When items pass out of stock as permanent issues, transfers or disposals, they become the responsibility of the plant record and inventory section, and are deleted from the stores list. Figure 6.3 shows a typical list for this type of stores. It is compiled in loose-leaf form to allow the insertion, extraction or substitution of additional sheets to keep the records correct.

CLASSIFICATION 3 - PLANT UNITS OR COMPOSITE ASSEMBLIES:-

TITLE PAGE:

| | |
|---|---|
| BINDER NO. | 1 |
| STORES MAIN CLASSIFICATION | 3/ |
| CATEGORY SUB-CLASSIFICATION | /01-05 |

PAGE 2:

EXPLANATORY NOTES WITH STOCK HEADING CODE AS SET OUT IN TEXT OF THIS CHAPTER.

RUNNING STOCK LIST PAGE:

RUNNING STOCK LIST

| RELATED COMPOSITE PLANT. | MAKE OF UNIT | MODEL OR TYPE | MAIN COMPONENTS OF COMPOSITE ASSEMBLY. | EXTRA FITTINGS INCLUDED. | ORDER NO. | DATE OF ORDER. | DATE RECEIVED | SOURCE IF REDUNDANT. |
|---|---|---|---|---|---|---|---|---|
| | | | | | | | | |

| STORES MAIN CLASSIFICATION | 3 | PLANT UNITS & COMPOSITE ASSEMBLIES |
|---|---|---|
| CATEGORY SUB-CLASSIFICATION | 04 | COMPRESSORS. |
| STOCK HEADING CODE | 01 | EMERGENCY STOCK |

| STORING LOCATION | | | | | PLANT NOS. ALLOCATED | LOANED | | RET'D | HELD FOR PROJECT | ISSUED | | REPLACEMENT ORDERED (DATE) | TRANSFERRED | | DISPOSED OF | |
|---|---|---|---|---|---|---|---|---|---|---|---|---|---|---|---|---|
| ACTIVITY | SECTION | BIN | AREA | CPD. | | TO | DATE | (DATE) | | TO | DATE | | TO | DATE | TO | DATE. |
| | | | | | | | | | | | | | | | | |

FIG. 6.3. Make-up of stores lists—classification 3.

## DETERMINING STOCK LEVELS

The choice of the stock levels for the various types of stores involves collaboration between the purchasing and stores department, the technical departments and the suppliers. The importance of the following factors should be appreciated.

Stocks must be kept to the lowest level consistent with both the foreseeable normal and emergency demands of production and with the delivery position. Inward deliveries should be so related to consumption and manufacturing times that live-stock conditions prevail where possible.

Stocks represent money tied up in the form of materials; these materials must be accounted for in cash terms. They are a part of the capital value of the undertaking on which interest charges must be taken into account.

They make further financial demands in provisions for storage and preservation from deterioration, damage and fire.

Some 'dead stock' is inevitable, when risks to production must be balanced against protracted deliveries, but such items must be separately justified. Any trend towards increased 'dead stock' must be guarded against.

In collaboration with the makers, close estimates of forward demand must be made; from these, the makers can build up their own stocks to give short-term delivery and reduce the need for the user to accommodate stocks from many suppliers on an undesirable scale.

Stocks should be centralised where possible, so that the minimum holding gives the maximum coverage.

Wherever possible, advantage should be taken of standardisation and interchangeability.

In the central workshops, unduly high standards of inspection and assessment for plant reconditioning must be avoided, and a high and economic scale of reclamation must be achieved.

The stock holdings should be based on maximum and minimum levels which allow for the operating conditions within the undertaking, the estimated requirements, past consumptions, and current delivery periods. The low level must allow demand to be met pending delivery of replenishments. The upper level should be such that the difference between it and the minimum makes the best use of the delivery period, to keep the total holding as low as possible at any time. At the same time, 'buyer-craft' should not be overlooked here, and the economics of buying in larger quantities should be examined. The minimum is the ordering point; the difference between the minimum and the maximum is the ordering quantity.

These principles apply to all types of stocks. More detailed reference is made to the forecasting of plant spares and components holdings in Chapter 4.

It has been said that prophecy is the most fortuitous form of error. It should be the aim to use all the knowledge and experience available, to ensure that holdings are more truly assessed, and less fortuitous.

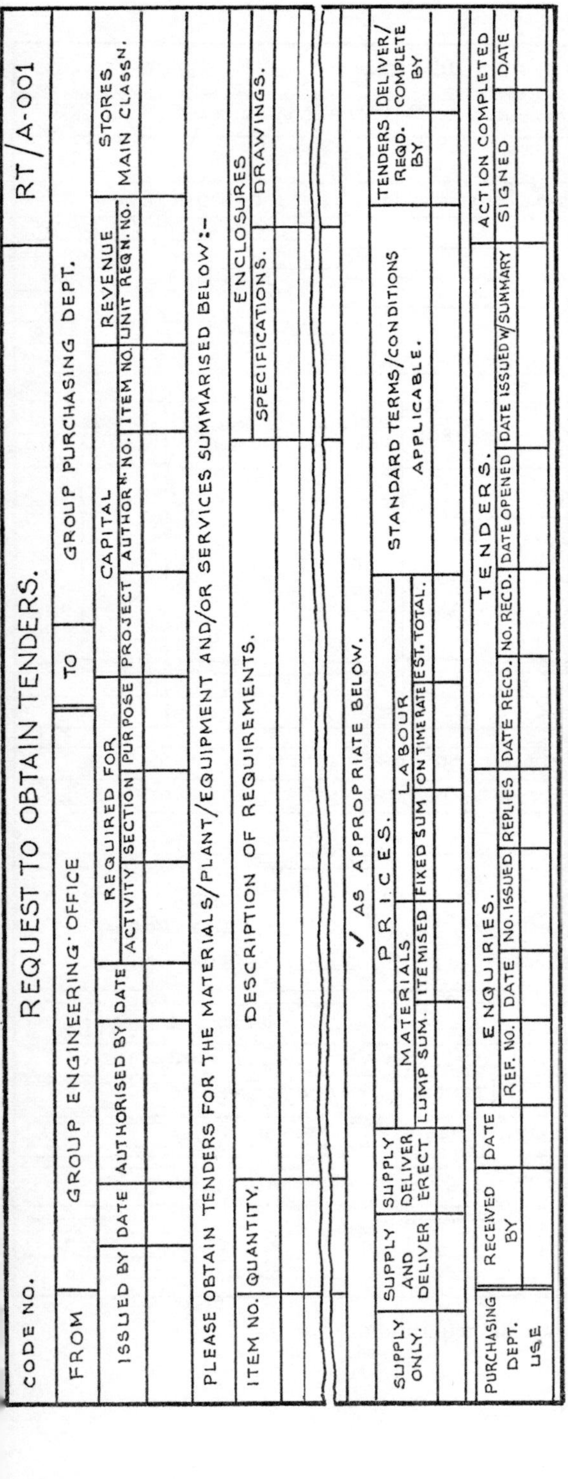


FIG. 6.4. Internal request to obtain tenders.

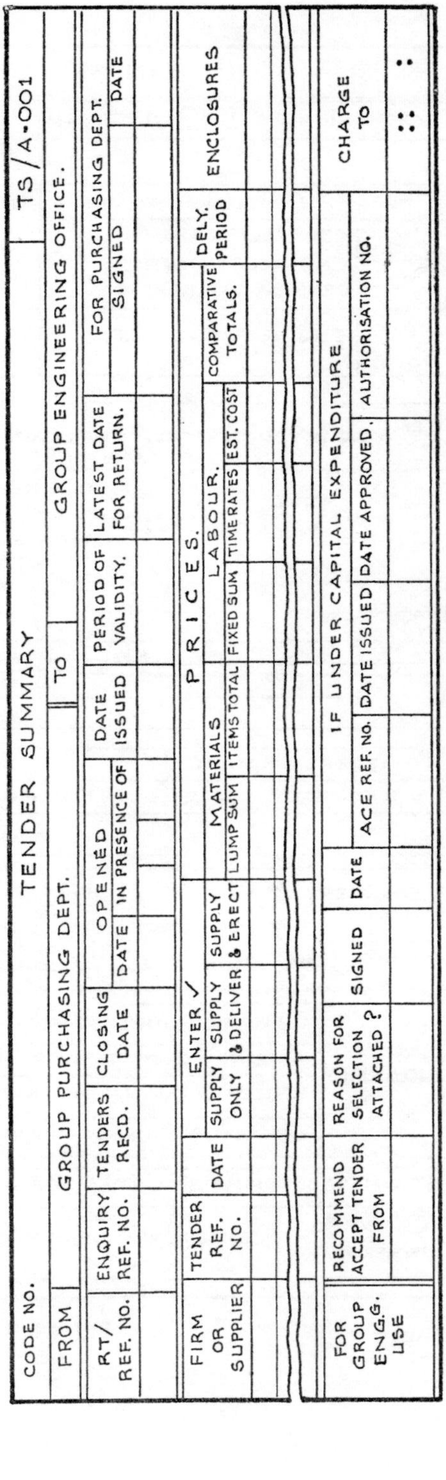


FIG. 6.5. Tender summary.

| CODE NO. | INTERNAL REVENUE REQUISITION | | | RR/A-001 |
|---|---|---|---|---|
| MECHANICAL ☐ | ELECTRICAL ☐ | COMBINED ☐ | STORES MAIN CLASS N. ☐ | |

| FROM | | | | | | TO | |
|---|---|---|---|---|---|---|---|
| ✓ WHERE APPROPRIATE | | ISSUED BY | DATE | APPROVED BY | DATE | PURCHASING DEPT. | |
| GROUP | CENTRAL STORES ☐ | | | | | RECEIVED BY | DATE |
| | ENGINEERING OFFICE ☐ | | | | | | |
| | CENTRAL WORKSHOPS. ☐ | | | | | CHECKED BY | DATE |
| UNIT ......... | ENGINEERING OFFICE. ☐ | | | | | ORDER NO. | DATE |

| ITEM NO. | QUANTITY | DESCRIPTION OF MATERIALS AND/OR SERVICES REQD. | MAKERS' PART NOS. | VOCABULARY NUMBER. | COST ACTUAL | COST ESTD. |
|---|---|---|---|---|---|---|
| | | ORDER HEADING<br>..............................<br>.............................. | | | | |
| | | | | | | |
| | | | | | | |
| | | | | | | |

| REFERS ENQUIRY | | SUPPLIER OR CONTRACTOR | | |
|---|---|---|---|---|
| REF. NO. | DATE | ITEM NOS. | SOLE MAKER | SUGGESTED |
| | | | | |

| DELIVERY AND PHASING INSTRUCTIONS | BY ROAD/RAIL TO ................................TAGGED ..................<br>COMMENCE ..............(DELIVERY) ..................(SITE WORK)<br>COMPLETE ..............(DELIVERY) ..................(SITE WORK) |
|---|---|

| ENCLOSURES | DRAWING NUMBERS. | | TOTAL NO. OF PRINTS ☐ |
|---|---|---|---|
| | ACCEPTED QUOTATION | FROM .................... REF. NO. .......... | DATE .......... |
| | REJECTED QUOTATIONS (NUMBER).... ☐ | TENDER SUMMARY REF. NO. TS/ | |
| | CENTRAL W'SHOPS PRE-PRINTED PARTS LISTS | SHEETS ☐ | JOB SERIAL NO. ☐ |

STANDARD TERMS & CONDITIONS APPLICABLE:-

| CHARGING INSTRUCTION | CENTRAL W'SHOPS JOB SERIAL NO. ☐ | ACTIVITY COST CODE :: : |
|---|---|---|

FIG. 6.6. Internal revenue requisition.

| CODE NO. | INTERNAL CAPITAL REQUISITION | CR/A-001 |
|---|---|---|

| FROM | | | | | | | TO | | | | | | |
|---|---|---|---|---|---|---|---|---|---|---|---|---|---|
| GROUP ENGINEERING OFFICE. | | | | | | | PURCHASING & STORES DEPT. | | | | | | |
| ISSUED BY | DATE | APPROVED BY | DATE | AUTHORISATION | | | RECEIVED BY | DATE | CHECKED BY | DATE | ORDER OR CONTRACT No. | | DATE |
| | | | | ACC. REF. | AUTH$^{N}$. No. | | | | | | | | |
| | | | | | | | | | | | | | |

| ITEM No. | QUANTITY. | DESCRIPTION OF MATERIALS AND/OR SERVICES REQD. | ITEMISED PRICES | | |
|---|---|---|---|---|---|
| | | | MATERIALS | LABOUR | TOTAL |
| | | ORDER OR CONTRACT HEADING<br>------------------------------<br>------------------------------ | | | |
| | | | | | |
| | | | | | |
| | | | | | |
| | | | | | |
| | | TOTALS. | | | |

| REFERS ENQUIRY | | SUPPLIER OR CONTRACTOR | | |
|---|---|---|---|---|
| REF NO. | DATE | NAME AND ADDRESS | TENDER REF. NO. | DATE |
| | | | | |

| DELIVERY AND PHASING INSTRUCTIONS | BY ROAD/RAIL TO ........................ TAGGED ..................<br>COMMENCE ................(DELIVERY)......................(SITE-WORK)<br>COMPLETE ................(DELIVERY)......................(SITE-WORK) |
|---|---|

| ENCLOSURES | DRAWING NOS: | TOTAL NO. OF PRINTS ☐ |
|---|---|---|
| | ACCEPTED TENDER DOCUMENTS: | |
| | REJECTED TENDER DOCUMENTS....SETS ☐ | TENDER SUMMARY REF. NO. TS/ |
| | DETAILED SPECIFICATION ON ADDITIONAL SHEETS NUMBERED | TO |

STANDARD TERMS & CONDITIONS APPLICABLE :-

| CONTACTS | | NAME | DESIGNATION | TEL. NO. |
|---|---|---|---|---|
| | FOR COMMERCIAL ASPECTS... | | | |
| | FOR TECHNICAL ASPECTS..... | | | |

| CHARGING INSTRUCTIONS. | | | | | |
|---|---|---|---|---|---|
| STORES MAIN CLASS$^{N}$. | CENTRAL W'SHOPS JOB SERIAL NO. | ACTIVITY COSTING CODE | PROJECT | | |
| | | | AUTH$^{N}$. NO. | SECTION | ITEM |
| | | :: : | | | |

FIG. 6.7. Internal capital requisition.

| APPLICATION FOR PLANT/EQUIPMENT/SERVICES UNDER CAPITAL EXPENDITURE | | | | | | | | ACE/A-001 |
|---|---|---|---|---|---|---|---|---|
| REQUIRED FOR | | | REASON IF ADDITIONAL | DETAILS OF EXISTING PLANT IF REPLACEMENT. | | | | |
| ACTIVITY | SECTION | PURPOSE | | PLANT NO. | DESCRIPTION | REASON | PLANT DISPOSAL AUTH. NO. IF ISSUED. | |
| | | | | | | | | |
| PROJECT | AUTH. NO. | SECTION/ITEM | | | | | | |
| | | | | | | | | |

FOR PURCHASE NEW:

| ITEM NO. | NO. OR Q'TITY. | DESCRIPTION | CLASSIFICATION CODES | | T | E | AMOUNT | | | | | | | DELY. | FOR STOCK | | | FOR DIRECT USE | |
|---|---|---|---|---|---|---|---|---|---|---|---|---|---|---|---|---|---|---|---|
| | | | | | | | EXTERNAL | | | INTERNAL | | | TOTAL | PROMISE | DATE REQD. | PRESENT STOCK | ISSUE RATE | INSTAL DATE | DATE TO PUT TO USE |
| | | | STORES | FINANCE | | | MTLS. | LAB. | TOTAL | MTLS. | LAB. | TOTAL | | | | | | | |
| | | | | | | | | | | | | | | | | | | | |
| | | | | | | | | | | | | | | | | | | | |

T - SELECTED TENDER PRICE. TENDER SUMMARY WITH APPRAISAL ATTACHED. E - ESTIMATED PRICE.

FOR TRANSFER EX STOCK AS (1) NEW (2) RECONDITIONED (3) AVAILABLE AS REDUNDANT:

| ITEM NO. | NO. OR Q'TITY. | DESCRIPTION | FOR PLANT RECORDS | | | | | | FOR STORES. | | | | | | | |
|---|---|---|---|---|---|---|---|---|---|---|---|---|---|---|---|---|
| | | | AVAILABLE AS REDUNDANT. | | | | | | AVAILABLE EX STOCK. | | | | | | | IS STOCK MAKE-UP NECESSARY? |
| | | | PLANT NO. | WHERE LOCATED | AGE | CONDITION | BOOK VALUE | ENT D. IN WEEKLY SUMMARY NO. | PLANT NO. | WHERE LOCATED | NEW OR S/HAND | MAIN CLASS[N]. | STORES CODING | BOOK VALUE | ENT'D IN WEEKLY SUMMARY No | |
| | | | | | | | | | | | | | | | | |
| | | | | | | | | | | | | | | | | |

IF ANSWER IN LAST COLUMN IS "YES", COMPLETE UPPER SECTION FOR NEW PURCHASE.

| | UNIT ACTIVITY | | GROUP HEADQUARTERS. | | | | | | | |
|---|---|---|---|---|---|---|---|---|---|---|
| | | | | | CAPITAL EXPENDITURE COMMITTEE. | | | | FINANCE | |
| | ENGINEERING | PRODUCTION | ENGINEERING | PRODUCTION | TRANSFER APPROVED | ORDERING APPROVED | REJECTED | REMARKS. AND/OR INSTRUCTIONS | AUTHORISATION NO. | CHARGING CODE |
| SIGNED | | | | | | | | | | :: : |
| DATE | | | | | | | | | | |

FIG. 6.8. Application for plant/equipment/services under capital expenditure.

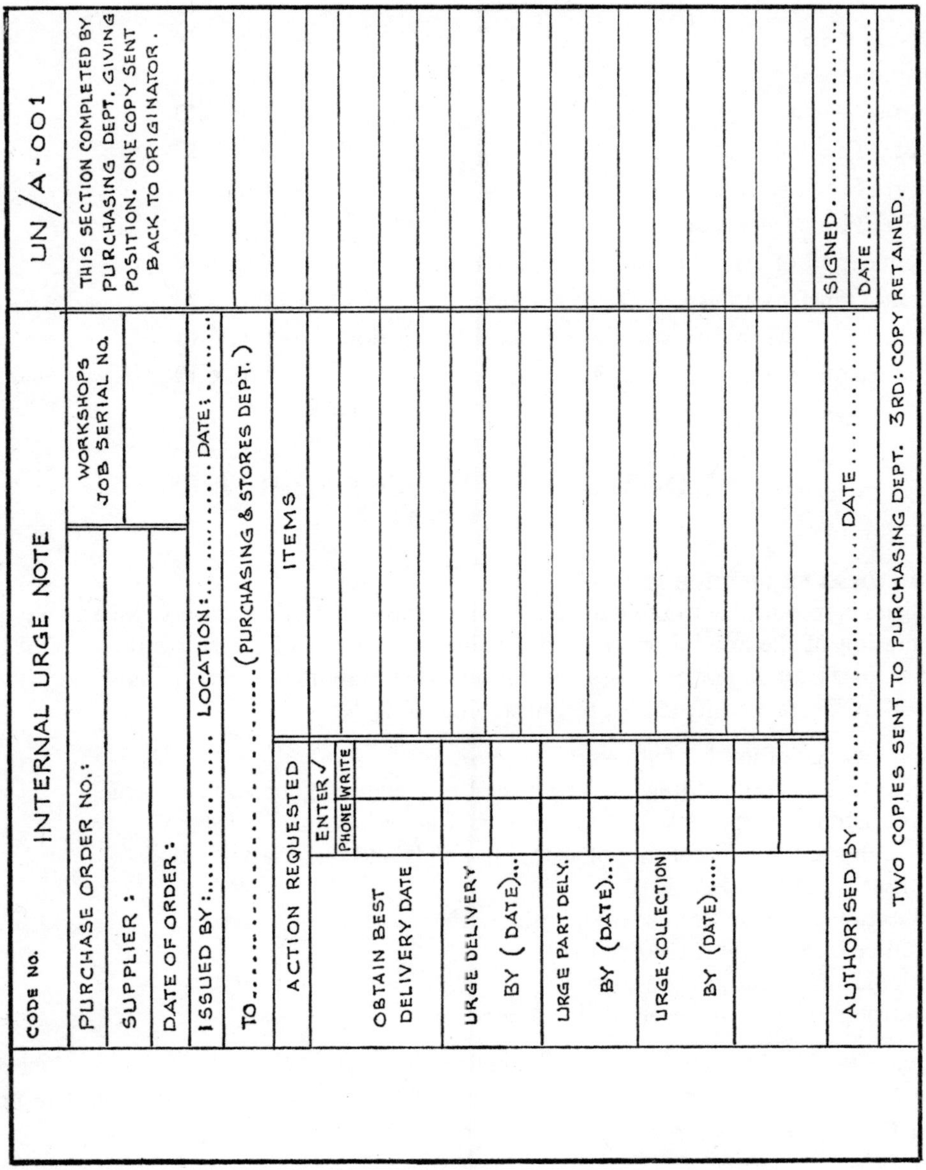

CODE No.

INTERNAL URGE NOTE

UN/A-001

PURCHASE ORDER No.:

WORKSHOPS JOB SERIAL No.

THIS SECTION COMPLETED BY PURCHASING DEPT. GIVING POSITION. ONE COPY SENT BACK TO ORIGINATOR.

SUPPLIER:

DATE OF ORDER:

ISSUED BY:............ LOCATION:............ DATE:........

TO.......................(PURCHASING & STORES DEPT.)

| ACTION REQUESTED | ENTER ✓ PHONE | ENTER ✓ WRITE | ITEMS |
|---|---|---|---|
| OBTAIN BEST DELIVERY DATE | | | |
| URGE DELIVERY BY (DATE)... | | | |
| URGE PART DELY. BY (DATE)... | | | |
| URGE COLLECTION BY (DATE)..... | | | |

SIGNED..................

DATE....................

AUTHORISED BY......................DATE..............

TWO COPIES SENT TO PURCHASING DEPT. 3RD: COPY RETAINED.

FIG. 6.9. Internal urging note.

## STORES DOCUMENTATION

In addition to the typical purchasing documents already referred to, some further documents related to the maintenance stores are noted in the following:

| | | |
|---|---|---|
| Goods Received Note | — | Fig. 6.10 |
| Inspection Slip | — | Fig. 6.11 |
| Stores Warrant | — | Fig. 6.12 |
| Stores Credit Note | — | Fig. 6.13 |
| Stores Stock Record Card | — | Fig. 6.14 |
| Stores Purchase Record Card | — | Fig. 6.15 |
| Notification of Miscellaneous Scrap for Disposal | — | Fig. 6.16 |
| Plant Transfer or Disposal Authorisation | — | Fig. 6.17 |
| Plant and Materials Transfer or Disposal Instruction | — | Fig. 6.18 |
| Confirmation of Plant Disposal | — | Fig. 6.19 |

## STAFF DUTIES AND RESPONSIBILITIES (see Fig. 6.1)

### Group stores superintendent

This official is responsible to the purchasing and stores manager for the control and day-to-day operation of the administrative and functional activities of the stores throughout the Group. This is an important post, of considerable standing in any sizeable industrial organisation. The superintendent's responsibilities include:

Maintenance of levels of stocks.
Receipt, inspection, storage and recording of incoming goods and materials.
Safe custody and preservation of goods and materials during storage.
Issue, delivery and recording of outgoing goods and materials.
Utilisation of storage space.
Collection and disposal of redundant, obselete, and scrap materials.
Maintenance of stores records.
Periodic stocktaking.
Control and supervision of all stores personnel.

Efficiency in the performance of these functions has a marked effect on production, the services to production, and the financial results of any industrial undertaking. The post calls for organising ability of a high order, a vigilant nature, and complete integrity.

### Group stock control officer

This official is responsible to the Group stores superintendent for the stock control function of stores administration; this includes:

Fixing stock holdings in collaboration with the user departments, and their disposition at the various stores in the Group.
Maintenance and adjustment of these holdings.
Control of receipts, issues and loans.
Initiation of appropriate action in case of shortage, damage or loss.
Initiation of action in respect of redundant, obsolete and scrap stocks.
Control of purchases related to stock holdings and redundant materials available.
Periodic stocktaking.
Records and procedures appropriate to these functions.
Control of clerical staff.

### *Group value control officer*

This official is responsible to the Group stores superintendent for the value control function of stores administration; his duties include:

Establishment and recording of the values of stocks.
Maintenance and adjustment of these values.
Debiting and crediting stock movements to the appropriate charge accounts.
Establishing the valuation of proposed purchases where the normal enquiry procedure is not applied.
Periodic stock valuation.
Records and procedures appropriate to these functions.
Control of clerical staff.

### *Central storekeeper*

This official is responsible to the Group stores superintendent for the stores processes at Group level. His duties include:

Receipt, inspection and storage of incoming goods.
Safe custody and preservation of goods and materials during storage.
Location, transfer, issue, and consignment of outgoing goods and materials.
Handling of redundant, obsolete and scrap materials.
Utilisation of storage space.
Maintenance of the bin cards as a check on stock control records.
Collaboration in stocktaking.
Records and procedures appropriate to the foregoing.
Control and supervision of stores labour.

### *Unit storekeepers*

These persons are directly responsible to the Group stores superintendent. Their duties are equivalent to the latter on the Unit scale. Their subordinate staffs cover duties equivalent to their counterparts at Group level, on the Unit scale.

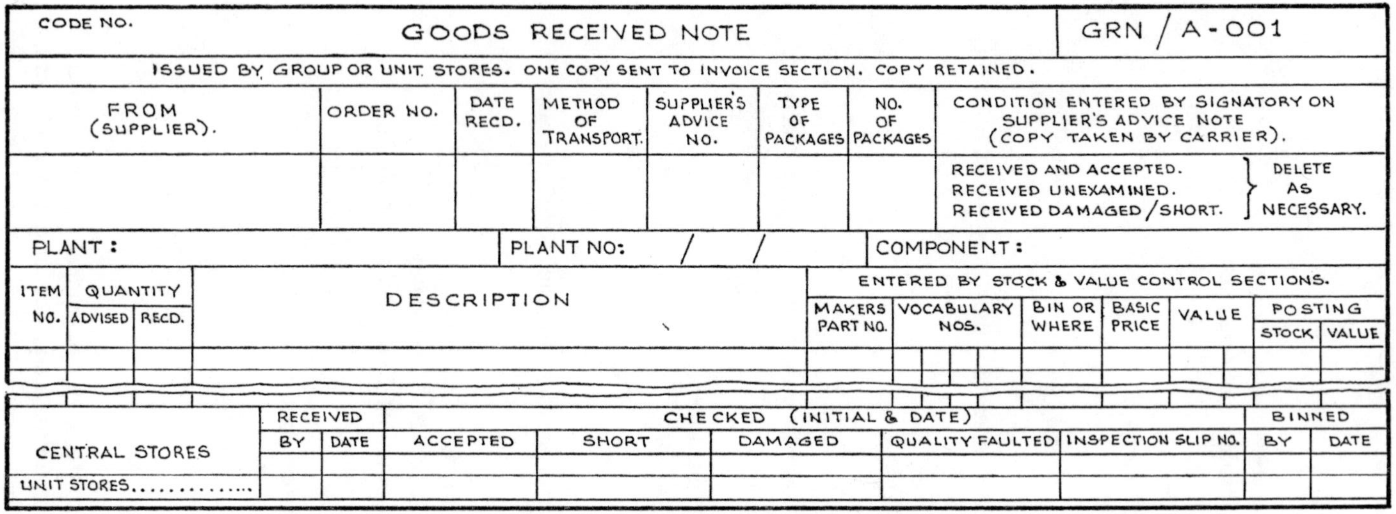

CODE NO. GOODS RECEIVED NOTE GRN / A-001

ISSUED BY GROUP OR UNIT STORES. ONE COPY SENT TO INVOICE SECTION. COPY RETAINED.

| FROM (SUPPLIER). | ORDER NO. | DATE RECD. | METHOD OF TRANSPORT. | SUPPLIER'S ADVICE NO. | TYPE OF PACKAGES | NO. OF PACKAGES | CONDITION ENTERED BY SIGNATORY ON SUPPLIER'S ADVICE NOTE (COPY TAKEN BY CARRIER). |
|---|---|---|---|---|---|---|---|
| | | | | | | | RECEIVED AND ACCEPTED. RECEIVED UNEXAMINED. RECEIVED DAMAGED/SHORT. DELETE AS NECESSARY. |

PLANT: PLANT NO: / / COMPONENT:

| ITEM NO. | QUANTITY ADVISED | QUANTITY RECD. | DESCRIPTION | MAKERS PART NO. | VOCABULARY NOS. | BIN OR WHERE | BASIC PRICE | VALUE | POSTING STOCK | POSTING VALUE |
|---|---|---|---|---|---|---|---|---|---|---|
| | | | | ENTERED BY STOCK & VALUE CONTROL SECTIONS. | | | | | | |

| | RECEIVED BY | RECEIVED DATE | CHECKED (INITIAL & DATE) ACCEPTED | SHORT | DAMAGED | QUALITY FAULTED | INSPECTION SLIP NO. | BINNED BY | BINNED DATE |
|---|---|---|---|---|---|---|---|---|---|
| CENTRAL STORES | | | | | | | | | |
| UNIT STORES............. | | | | | | | | | |

FIG. 6.10. Internal goods received note.

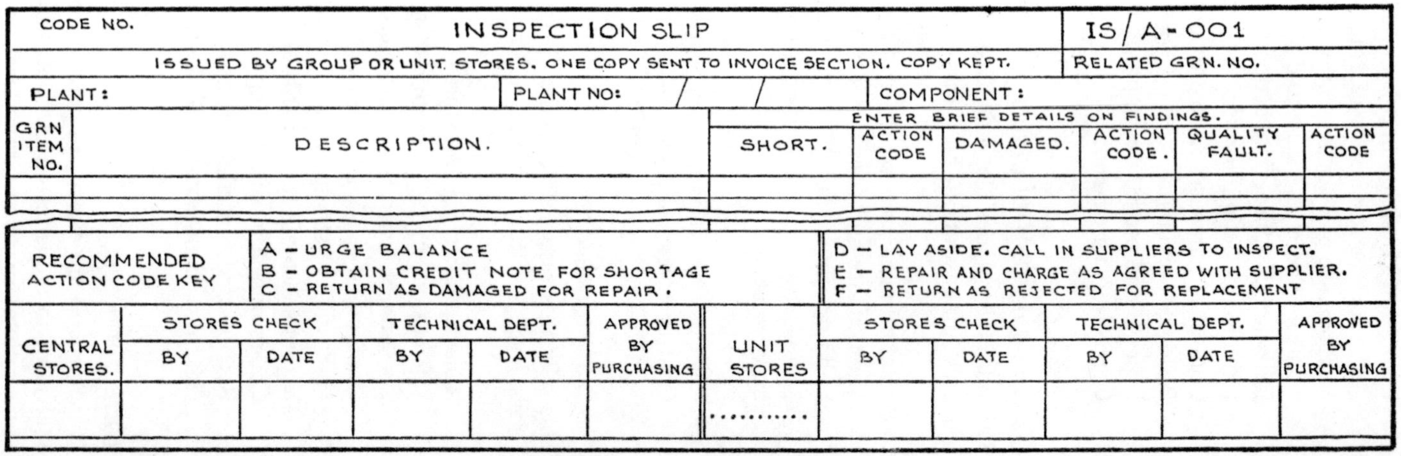

CODE NO. INSPECTION SLIP IS/A-001

ISSUED BY GROUP OR UNIT STORES. ONE COPY SENT TO INVOICE SECTION. COPY KEPT. RELATED GRN. NO.

PLANT: PLANT NO: / / COMPONENT:

| GRN ITEM NO. | DESCRIPTION. | SHORT. | ACTION CODE | DAMAGED. | ACTION CODE. | QUALITY FAULT. | ACTION CODE |
|---|---|---|---|---|---|---|---|
| | | ENTER BRIEF DETAILS ON FINDINGS. | | | | | |

RECOMMENDED ACTION CODE KEY

A – URGE BALANCE
B – OBTAIN CREDIT NOTE FOR SHORTAGE
C – RETURN AS DAMAGED FOR REPAIR.
D – LAY ASIDE. CALL IN SUPPLIERS TO INSPECT.
E – REPAIR AND CHARGE AS AGREED WITH SUPPLIER.
F – RETURN AS REJECTED FOR REPLACEMENT

| CENTRAL STORES. | STORES CHECK BY | STORES CHECK DATE | TECHNICAL DEPT. BY | TECHNICAL DEPT. DATE | APPROVED BY PURCHASING | UNIT STORES ........... | STORES CHECK BY | STORES CHECK DATE | TECHNICAL DEPT. BY | TECHNICAL DEPT. DATE | APPROVED BY PURCHASING |
|---|---|---|---|---|---|---|---|---|---|---|---|
| | | | | | | | | | | | |

FIG. 6.11. Inspection slip.

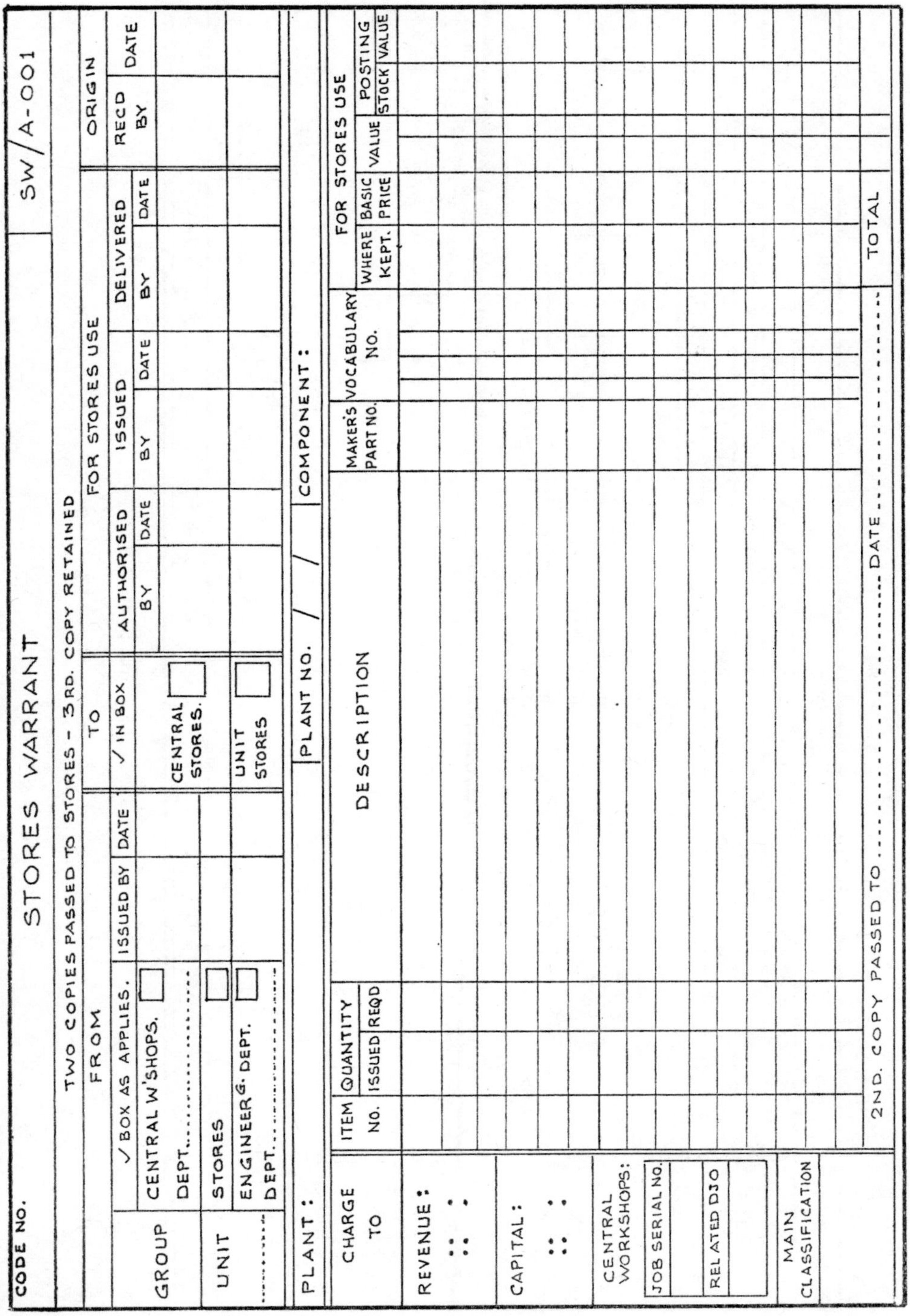


FIG. 6.12. Stores warrant.

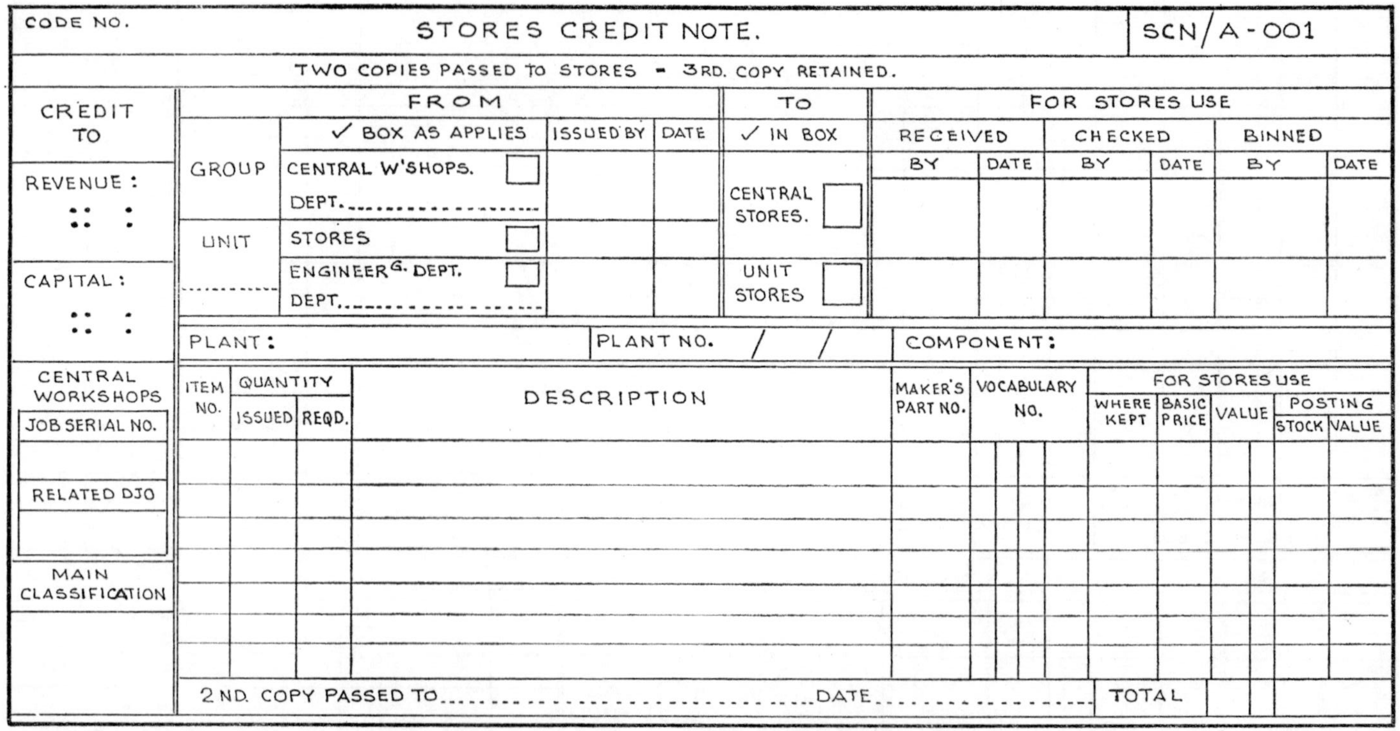


FIG. 6.13. Stores credit note.

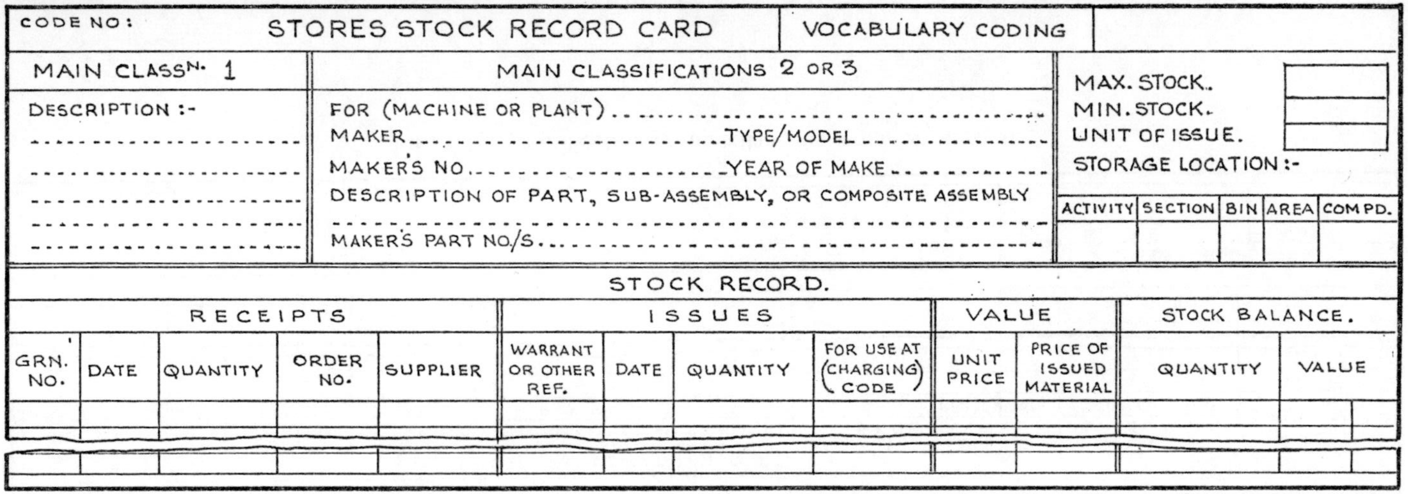


FIG. 6.14. Stores stock record card.

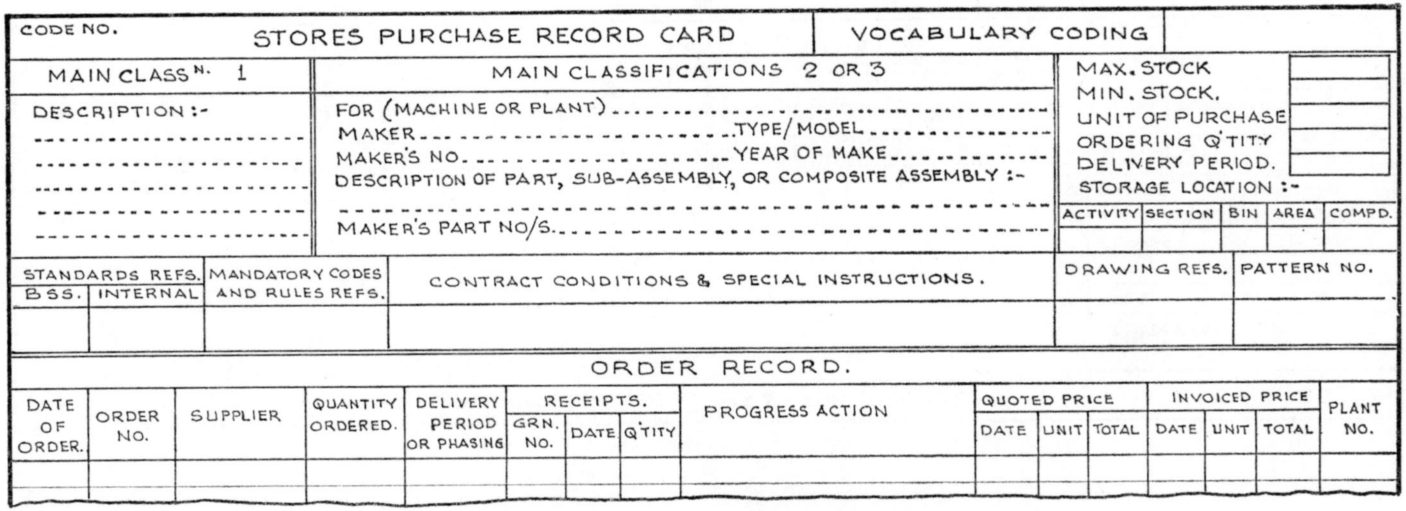


FIG. 6.15. Stores purchase record card.

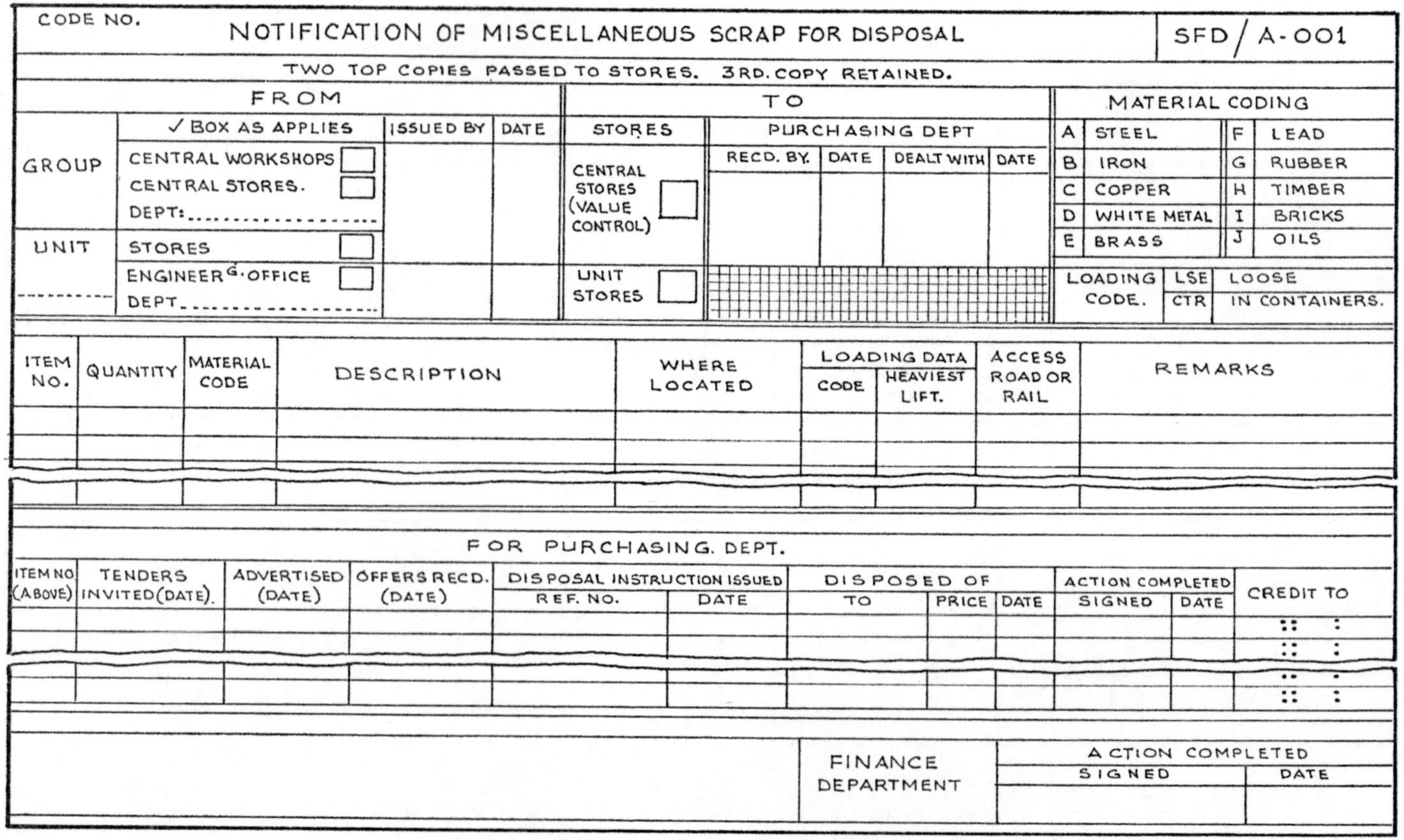
CODE NO. NOTIFICATION OF MISCELLANEOUS SCRAP FOR DISPOSAL SFD/A-001

TWO TOP COPIES PASSED TO STORES. 3RD. COPY RETAINED.

| FROM | | | | TO | | | | | MATERIAL CODING | | | |
|---|---|---|---|---|---|---|---|---|---|---|---|---|
| | ✓ BOX AS APPLIES | ISSUED BY | DATE | STORES | PURCHASING DEPT | | | | | | | |
| | | | | | RECD. BY. | DATE | DEALT WITH | DATE | A | STEEL | F | LEAD |
| GROUP | CENTRAL WORKSHOPS ☐ CENTRAL STORES. ☐ DEPT: .................. | | | CENTRAL STORES (VALUE CONTROL) ☐ | | | | | B | IRON | G | RUBBER |
| | | | | | | | | | C | COPPER | H | TIMBER |
| | | | | | | | | | D | WHITE METAL | I | BRICKS |
| UNIT | STORES ☐ | | | | | | | | E | BRASS | J | OILS |
| | ENGINEER[G]. OFFICE ☐ DEPT .................. | | | UNIT STORES ☐ | | | | | LOADING CODE. | LSE | LOOSE | |
| | | | | | | | | | | CTR | IN CONTAINERS. | |

| ITEM NO. | QUANTITY | MATERIAL CODE | DESCRIPTION | WHERE LOCATED | LOADING DATA | | ACCESS ROAD OR RAIL | REMARKS |
|---|---|---|---|---|---|---|---|---|
| | | | | | CODE | HEAVIEST LIFT. | | |
| | | | | | | | | |

FOR PURCHASING. DEPT.

| ITEM NO (ABOVE) | TENDERS INVITED (DATE). | ADVERTISED (DATE) | OFFERS RECD. (DATE) | DISPOSAL INSTRUCTION ISSUED | | DISPOSED OF | | | ACTION COMPLETED | | CREDIT TO |
|---|---|---|---|---|---|---|---|---|---|---|---|
| | | | | REF. NO. | DATE | TO | PRICE | DATE | SIGNED | DATE | |
| | | | | | | | | | | | :: : |

| | FINANCE DEPARTMENT | ACTION COMPLETED | |
|---|---|---|---|
| | | SIGNED | DATE |

FIG. 6.16. Notification of miscellaneous scrap for disposal.

| CODE NO. | PLANT TRANSFER OR DISPOSAL AUTHORISATION | | | | PT.DA/ |
|---|---|---|---|---|---|
| FROM | FOR CAPITAL EXPENDITURE COMMITEE | | FOR FINANCE DEPT. | | RELATED MOVEMENT ADVICE |
| | AUTHORISED BY | DATE | ISSUED BY | DATE | PMRA/ |

| ITEM NO. ON W.R. | NO. OR Q'TITY | NEW OR USED | DESCRIPTION | INTERNAL PLANT NO. | ACTION AUTHORISED | | | | | | | | ASSESSED CURRENT VALUE |
|---|---|---|---|---|---|---|---|---|---|---|---|---|---|
| | | | | | TRANSFER | | | | | | DISPOSE OF | | |
| | | | | | FROM | | | TO | | | | | |
| | | | | | STOCK AT | ACTIVITY | CODE | EXTERNAL PURCHASER | ACTIVITY | CODE | AS USABLE | AS SCRAP | |

| TO | PURCHASING & STORES DEPARTMENT | | | | | |
|---|---|---|---|---|---|---|
| | INSTRUCTIONS ISSUED BY | DATE | ISSUED TO | DATE | ACTION COMPLETED | DATE |
| | | | | | . (MANAGER) | |

| TRANSFERRED | PMRA TO PLANT RECORDS | TENDERS INVITED | ADVERTISED | OFFERS RECD. | INSTRUCTION TO STORES | | DISPOSED OF | | | CONFIRMED TO PLANT RECORDS | |
|---|---|---|---|---|---|---|---|---|---|---|---|
| DATE | (DATE) | DATE | DATE | DATE | REF. NO. | DATE | TO | PRICE | DATE | REF. NO. | DATE |

FIG. 6.17. Plant transfer or disposal authorisation.

| CODE NO. | PLANT AND MATERIALS TRANSFER OR DISPOSAL INSTRUCTION. | | | | | | | | TDI /A-001 |
|---|---|---|---|---|---|---|---|---|---|
| FROM | PURCHASING & STORES DEPT. | | | | TO | CENTRAL STORES | | | |
| | ISSUED BY | DATE | AUTHORISED BY | DATE | | RECEIVED BY | DATE | FOR ACTION TO | DATE |

PLANT:

| NEW OR USED | BRIEF DESCRIPTION | INTERNAL PLANT NO. | PDTA REF NO. | PDTA ITEM NO. | WHERE LOCATED | ACTION TO BE TAKEN | ACTION COMPLETED SIGNED | ACTION COMPLETED DATE | CONFIRMED TO PLANT RECORDS REF. NO. | CONFIRMED TO PLANT RECORDS DATE |
|---|---|---|---|---|---|---|---|---|---|---|
| | | | | | | | | | | |

MISCELLANEOUS SCRAP:

| Q'TITY. | MATERIAL | BRIEF DESCRIPTION | SFD REF NO. | SFD ITEM NO. | WHERE LOCATED | ACTION TO BE TAKEN | LOADING DATA | ACCESS DATA | ACTION COMPLETED SIGNED | ACTION COMPLETED DATE |
|---|---|---|---|---|---|---|---|---|---|---|
| | | | | | | | | | | |

FIG. 6.18. Plant and materials transfer or disposal instruction.

| CODE NO. | CONFIRMATION OF PLANT DISPOSAL | | | | | | | | CD/A-001 |
|---|---|---|---|---|---|---|---|---|---|
| FROM | CENTRAL STORES | | | | TO | GROUP PLANT RECORDS & INVENTORY | | | |
| | ISSUED BY | DATE | AUTHORISED BY | DATE | | RECEIVED BY | DATE | ACTION COMPLETED BY | DATE |

THE UNDERNOTED PLANT ITEMS HAVE NOW BEEN DISPOSED OF. ENTER IN YOUR NEXT "WEEKLY RETURN OF PLANT MOVEMENT & RECEIPT" TO FINANCE. EXTRACT CARDS & DESTROY. MARK ITEMS IN INVENTORY AS "DISPOSED OF": -

ENTER IF APPLICABLE — COMPOSITE PLANT OF WHICH ITEMS FORM A PART

LOCATION: ........ PLANT: ........ SECTION: ........

| ITEM NO. | BRIEF DESCRIPTION | INTERNAL PLANT NO. | LAST LOCATION | DISPOSAL AUTHORISATION (PDTA) REF. NO. | DISPOSAL AUTHORISATION (PDTA) DATE | FOR PLANT RECORDS: DISPOSAL ENTD. ON W.R. REF. NO. | DISPOSAL ENTD. ON W.R. DATE. | INVENTORY CORRECTED (DATE) | CARDS DESTROYED (DATE) | NOTIFIED (DATE) DRG. OFFICE | NOTIFIED (DATE) MAINT. CONTROL |
|---|---|---|---|---|---|---|---|---|---|---|---|
| | | | | | | | | | | | |

FIG. 6.19. Confirmation of plant disposal.

# STORES ACCOMMODATION, LAY-OUT AND FACILITIES

The provision of suitable stores accommodation, lay-out and facilities includes collaboration between the purchasing and stores department and the technical departments to be served. The following are basic factors which require joint study:

## The nature of the various categories of stores involved

This is discussed earlier in this Chapter.

## The general pattern of holdings and distribution throughout the Group

This depends upon the nature of the undertaking, and also on geographical considerations. In general, supplies should be kept as near as possible to their point of use, to minimise handling and transport charges and to give prompt service. In a small works or factory, of compact lay-out, a central store, suitably located, may well be suitable. In a large works or factory, central stores with sub-stores located at various sections of the works are usually adopted. In the type of industrial undertaking which is considered in this book, this latter principle of decentralisation is essential. It is most satisfactorily organised on the basis that a central stores is located at the Group's headquarters, closely associated with the Group's purchasing activity, and allied to the central workshops; regular bulk or quota issues, general stores and sundries replenishments, as well as plant, spares and components, are distributed to the Units' stores. For general stores and sundries, and the common spares and components supplied from the central stores, the Units' stocks should be kept as low as possible, with bulk issues at short time intervals, in preference to increasing the Units' holdings and delivering at longer intervals.

Further segregation is required to complete the pattern of stores holdings and distribution throughout the Group. The locations of the stocks require more detailed description; that is, their positions in the stores proper in bins, racks, shelves or floor areas; in open compounds; or at specific points near the place of use. In compiling the lists already discussed, this segregation is provided for. This paves the way for the consideration of stores accommodation, lay-out and space requirements.

## The basic lay-out of accommodation at Group level, and at the Units

This is derived from the various stages of stores operation—that is: receipt, inspection, storage and issue—and relating these stages to the various types of stores involved, and the general pattern of holdings and distribution already discussed. The basic requirements may be summarised as follows:

*See over for Basic Requirements*

| Type of Stores | Type of Accommodation or Storage Provision | Storage Category |
|---|---|---|
| Loose small-issue items of packeted delivery type | Trays or drawers incorporated in bins framework | Internal binned storage |
| Small items stocked in quantity | Troughed bins | |
| Medium items stocked in quantity | Shelved bins partitioned to suit | |
| Larger supplies and materials requiring covered storage | Designated floor spaces, or defined bays, supported on pallets or racks when advantageous. Provision of handling facilities (cranes, electric trucks, forklifts, conveyors) to consider | Internal bulk storage |
| Items suitable for open storage, with protection treatment if required | Open compounds with boundary fences. Provision of handling facilities (cranes, electric trucks, forklifts) to consider | Open fenced compound storage |
| Heavy items peculiar to a particular location, which involve special handling and transport, which could cause production loss in emergency | Selected site or area, as near to point of use as possible. Handling facilities to consider | Selected site storage |

The following points should be noted:

Materials and supplies, wherever stored, should pass through check points during receipt, inspection or issue.

Open compounds should be securely fenced, and the routing of materials in and out should facilitate control at the check points.

Road and rail transport services should be conveniently located.

Provision should be made for speedy handling of receipts, movements and issues. At receipt and dispatch points, there should be quays or platforms at the floor level of the wagon or vehicle, to simplify handling. Holding areas should be provided for unpacking, inspection and repacking. Handling facilities should be provided to reduce labour effort and travel; these may include:

Gantry or transporter cranes in stockyards.

Jib cranes at loading or unloading points.

Overhead travelling cranes in the heavy plant section of the stores.

| PURPOSE | SPACE REQUIREMENTS IN SQ. FT. | | | | | | | | | | | | | |
|---|---|---|---|---|---|---|---|---|---|---|---|---|---|---|
| | STORES. | | | | | | | | | | SELECTED SITE | | | |
| | INTERNAL | | | | | | EXTERNAL | | | | | | | |
| | BINNED STORAGE | | BULK STORAGE | | OTHER FLOOR SPACE | | OPEN FENCED COMPD. | | OTHER GROUND SPACE | | INTERNAL | | EXTERNAL | |
| | INITIAL | FUTURE | INITIAL | FUTURE | INITIAL | FUTURE | INITIAL | FUTURE | INITIAL | FUTURE | INITIAL | FUTURE | INITIAL | FUTURE |
| STORED COMMODITIES AND EQUIPMENT. | | + | | + | | + | | + | | + | | + | | + |
| TURN-ROUND AND HOLDING AREAS. | | | | | | | | | | | | | | |
| ROAD AND RAIL ACCESS. | | | | | | | | | | | | | | |
| SAWING, CUTTING AND MEASURING-OUT. | | | | | | | | | | | | | | |
| AISLES AND GANGWAYS. | | | | | | | | | | | | | | |
| IDLE MOBILE TRANSPORT. | | | | | | | | | | | | | | |
| HANDLING DEVICES E.G.:- ROLLER CONVEYORS. | | | | | | | | | | | | | | |
| ADMINISTRATION AND CONTROL | | | | | | | | | | | | | | |
| TOTALS. | | | | | | | | | | | | | | |
| | | | | | | | | | | | | | | |

FIG. 6.20. Storage space assessment.

Electric trucks and trailers for access along the main aisle of internal and external stores, with use of pallets when advantageous.
Fork-lift trucks for main aisle access to tiered bin, shelf and rack storage, with use of pallets.
Roller conveyors for light point-to-point transport.

Space for waiting transport should be provided. Ready accessibility to all storage areas should be maintained, with adequate room and suitable floors to support the heaviest materials. Regular issues, or faster-moving items, should be those most easily reached. Travel within the confines of the stores and compounds should be kept to the minimum. Aisles and storage locations should be clearly marked with the appropriate stores classifications and codings.

Space should be fully used. Double or treble deck storage should be provided for binned light stores, with a small communication hoist between floors. Heavier items and equipment should be stored on the ground with an overhead travelling crane. Height should be used to stack appropriate types of medium bulk materials on tiered racks or pallets. Materials suitable for external storage should be so accommodated. Aisle widths should be related to the type of stores involved.

Particular provision should be made for fragile stores; for those with a high fire risk, such as petrol, fuel oil, oils and greases; for perishable stores; for stores with special risks, such as acids; for stores with a design or shape rendering them particularly prone to damage; and for those requiring protection against deterioration due to temperature changes or dampness.

Allowance should be made for future expansion, should the need arise.

Protection and preservation of stocks from theft, weather and fire should be assured.

The buildings should have good natural lighting, and adequate heating and ventilation.

## Assessment of the space requirements

This involves a study of the commodities and equipment to be stored. The space might be divided as shown in Fig. 6.20. To complete the assessment, a further breakdown of needs under each of the purposes shown is required.

## Locating the stores buildings, compounds and other holding points

For both the Group's stores and each Unit's stores, the foregoing considerations apply, together with their integration into the conditions at the various locations.

## Design of the building and compounds

The layout and facilities at these and other holding points involves the development and correlation of the features discussed in the relevant sub-sections of this Chapter.

*Chapter 7*

# TRANSPORT AND COMMUNICATIONS
# and
# Their Contribution to the Maintenance Function

Transport and communications play a vital part in present-day industrial organisation, and activity. They do not always receive the attention they merit, and indeed are often left to develop haphazardly. They are inevitably linked, and together, if well organised and integrated into the general administrative structure, can make a substantial contribution to the efficiency of the maintenance service. Only those aspects which relate to the engineering maintenance function, and contribute to its satisfactory fulfilment, are considered here.

## TRANSPORT

Transport is the movement of men and materials, which is important in engineering maintenance. It is not necessarily under the control of the engineering establishment; indeed, it may well include road, rail, sea and air traffic, depending on the nature and size of the undertaking, and the scale of transport requirements.

For maintenance purposes the movement of men and materials must be achieved with efficiency and economy. Attention is confined to road transport, which is predominant in most cases of internal service maintenance.

In smaller undertakings the responsibility for transport may be delegated to outside resources, either on a contract or ad-hoc basis, with suitable control arrangements. Alternatively limited internal facilities may be provided, augmented as required by external resources.

In larger undertakings, it is more usual for a transport section to be set up as a pool service, with its own fleet of vehicles and facilities to maintain them in satisfactory condition. Only occasional specialised requirements (such as heavy or large loads calling for low-loaders, or considerations of distance) involve the use of outside resources.

Where internal vehicles and facilities are employed, their operation and deployment may be controlled by the transport section, with maintenance as the responsibility of the maintenance engineering department, in common with the other fixed assets of the undertaking. Such arrangements are described in Chapter 5. On the other hand, maintenance may be contracted out on an organised basis, in which case the costs must be recognised and distributed as Company policy dictates. This is described in Chapter 8.

An alternative to a centralised transport service in larger undertakings is the allocation of vehicles to particular departments, under their own control and deployment; for example, the central stores service, the central workshops and the Units. This may have some merit in certain circumstances, but the economics of such a policy should be weighed against those of a centralised service, with a suitable range of vehicles and facilities, continuously deployed in the interests of the undertaking as a whole.

## THE RANGE OF TRANSPORT DEMANDS AND FACILITIES

The nature and range of transport demands and the facilities required to meet them vary widely, of course, with the type and organisation of the undertaking concerned. For maintenance purposes, the following are typical:

### For the stores service

Distribution of general stores, spares and components from central stores to central workshops and Units. This may involve scheduled regular deliveries to maintain 'buffer' stocks at the Units, plus ad-hoc runs in case of emergency.

Issue of transportable plant in the form of plant units or composite assemblies, to meet demands at the Units.

Collection of surplus issues for return to stock holding points.

Collection and disposal of redundant, obsolete and scrap materials and plant.

### For the central workshops

Transport of completed jobs and plant to the Units, including those under the periodic overhaul programme.

Transport of components, plant and equipment to outside firms employed to augment internal resources.

Transport of labour to and from site repair and installation work at the Units.

### For the production units or activities

Transport of components, plant and equipment to central workshops for processing, repair or overhaul, including those under the periodic overhaul programme.

Transport of components, plant and equipment to outside firms.

### General

The arrangements should be such that transport requirements to meet genuine unforeseen emergencies are promptly satisfied at any time, day or night, in addition to the organised transport to cover the foregoing demands.

The facilities and organisation to meet these demands should receive close attention; the nature of the materials to be handled, speed and convenience in loading and unloading, and the most effective deployment of vehicles and personnel should all be considered, so as to achieve efficiency with economy.

Men should be conveyed in reasonable comfort, protected from the weather, and their travelling time kept to a minimum. Scheduled routes may be operated to established time-tables, where these are beneficial. The use of containers for general stores and sundries, and the smaller types of spares and components, may be advantageous. Other features which contribute to rapid handling, labour-saving, and operational efficiency include:

Hydraulically-operated loading arms of up to 2 or 3 tons capacity incorporated in the vehicle design; these enable the driver to do many loading and unloading operations single-handed.

Loading bays or quays at lorry deck level, at such places as stores buildings, to facilitate the transfer of materials.

Gantries with electric lifting devices at key points, under which vehicles can run.

The size, type and range of vehicles and facilities must be selected to suit the circumstances existing. Standardisation within the fleet leads to economic maintenance. The range also merits close study. Discussions with the makers help to secure the advantage of the latest developments, contributing to low running costs and economic maintenance.

A chart illustrating the types of vehicles in use, and other facilities such as low-loaders, giving their main features, principal particulars (type and inside dimensions of body, height to floor or platform, wheel base, overall width, overall length, turning circle, ground clearance, tare weight and pay-load capacity) should be kept available and up to date. This is useful when considerations of route and access to site locations arise. The use of fleet hire facilities should also be considered.

## DOCUMENTATION

An efficient transport service can only be achieved if the user departments discipline their own demands. Their requirements should be clearly indicated in good time. Requests for transport should be in writing on a standard document giving all necessary data. Emergency calls should be genuine, and however they are initiated or actioned, the written request with all information should reach the transport section as quickly as possible. The spurious use of the term 'urgent' must be avoided if a good standard of service is to be maintained.

It is further desirable that the document used for intimating transport needs should confirm the nature and quantity of materials transported, with dates and signatures from the dispatch and receipt points.

Figure 7.1 illustrates a form of transport advice which meets these requirements. It is made out in triplicate at the point of origin, the two top copies being sent to the transport office and the third being retained for record purposes. The transport office passes one copy to the vehicle driver for obtaining signatures confirming loading and delivery, and retains the other for reference. The driver's copy is returned to the transport office on completion of the job. This puts the transport office in a position to answer any queries, and to investigate any complaints of non-delivery, delay, or mis-direction. This document is referred to under the documentation and procedures for central workshops in Chapter 4.

CODE NO.

**MEN AND MATERIAL TRANSPORT ADVICE**

TA. NO............

| ISSUED BY................................ | ACTIVITY/SECTION ........................ | DATE ............ |
|---|---|---|

TWO TOP COPIES TO TRANSPORT OFFICE. 3RD. COPY RETAINED.

MATERIALS.

PLANT........................................ PLANT NO..../..../...... COMPONENT........................................

| COLLECT FROM | NO. OR QUANTITY. | DESCRIPTION. | DELIVER TO | WORKS ORDER DATED | CHARGING CODE |
|---|---|---|---|---|---|
| | | | | | :: : |
| | | | | | :: : |
| | | | | | :: : |
| | | | | | :: : |
| | | | | | :: : |
| | | | | | :: : |

URGENCY (ENTER AND/OR DELETE AS REQUIRED):

| COLLECT | NOW READY | AT.......... AM/PM ON.............. | WILL ADVISE WHEN READY |
|---|---|---|---|
| DELIVER | DIRECT/URGENT | NOT LATER THAN ................. | ON CONVENIENT DATE TO BE AGREED. |

HANDED OVER BY........................................ DATE DELIVERED........................................
UPLIFTED BY........................................ RECEIVED BY........................................

MEN.

| COLLECT FROM | NO. | DATE | TIME | TRANSPORT TO | PICK-UP TIME | RETURN TO | WORKS ORDER DATED | CHARGING CODE |
|---|---|---|---|---|---|---|---|---|
| | | | | | | | | :: : |

VEHICLE................. DRIVER ........................ COMPLETED.................. (TRANSPORT OFFICER)

FIG. 7.1. Men and material transport advice.

# RECORDS

Records that the transport office can usefully keep for convenience and as aids to operational efficiency include:

## For messages

A record or log book of all messages left or received, and the action taken; for example:

| Date | Time | Rec'd from | Sent to | Message | Transport advice | | | | Action taken | By | Remarks |
|---|---|---|---|---|---|---|---|---|---|---|---|
| | | | | | To follow | Rec'd | Advice no. | Date | | | |
| | | | | | | | | | | | |

This is principally a record of breakdown demands or emergency action called for at short notice. When such demands become too frequent, or are unjustified, the situation should be discussed with the responsible persons at the point of origin. The columns relating to the transport advice identify any sources of demand which are ignoring the standard procedure.

## For transport demands

Transport advice notes can be segregated into four classes, namely: received and awaiting action (arranged in order of priority); examined and awaiting further information or instruction; in progress; and action completed.

## For vehicle maintenance

A suitable maintenance procedure, based on mileage, is described in Chapter 5. The records relating to mileage build-up and timely notification to the maintenance department for vehicle attention, are an essential part of the system.

## For operational costs

Periodic operating cost sheets and returns should be filed for reference, and used in the control of costs.

## For scheduled runs

A time-table of scheduled regular runs and routes should be displayed, so that they may be used to the best advantage.

## For mileage chart

A mileage chart covering the whole operational area, and constructed in the way commonly used and shown overleaf, can be displayed and employed with advantage. This can usefully include any mileage to external firms which are regularly visited.

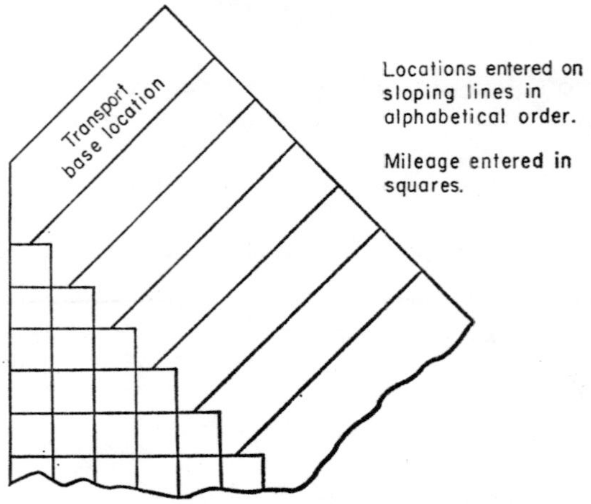


## For maps

A map of the locations of firms involved in transport, and the best route, together with their telephone numbers, is a further useful provision.

# COMMUNICATIONS

The term communications used to imply the mere transmitting of information between people, or places. However, the complexity of modern industry, mechanisation and automation, and the trend towards larger units and undertakings, have demanded more complex administrative structures and organisations to manage and control them. In these circumstances, communications have become both more important and more difficult. Good communications are essential to sound management, and to the attainment of efficiency.

Nowadays, communications are concerned not only with the passing of information and instructions within the undertaking, but also with the specialised instrumentation and communication systems needed for mechanisation and automation of production. Attention is here concentrated on those aspects relating to maintenance management and activity.

## Communication methods and systems

### *Written communications*

The organisation covering this aspect of communications is generally centred in a post and mailing section within the secretariat. It is concerned with the handling of inward and outward mail; internal communications between departments and sections; typing pool service; central filing; dissemination of administrative and technical literature; telex; and some aspects of computer terminals.

The routine procedures should be suited to the necessary level of service efficiency. Internal communications call for more frequent activity than those routed through the post, where postal and delivery times are the governing factor. A clear-cut means of identifying and segregating communications calling for priority treatment is desirable.

For internal communications, an unsealed transit envelope in a suitable range of sizes, is a simple device, affording worth-while economies.

An important factor, often overlooked, is the quoting of the references of both sender and addressee. This speeds the sorting and directing of correspondence, and also helps in filing. Trade journals, publications, advertising data, and so on should be directed on a selective basis, so that they reach the persons most likely to find them of interest in the shortest space of time. A gummed circulation strip may be attached by the secretariat; it should include provision for any person on the circulation list to draw the attention of others to a particular item or article, or to direct the publication to persons not listed. While these points may seem trivial, their cumulative effect is by no means insignificant.

## *Oral and visual communications*

The following gives some indication of the range of systems available:

The national telephone system (GPO).

Internal telephone systems:

> Automatic dialling systems; push button systems; the magneto system; the guided-wire radio telephone system; the radio telephone system; the paging inductive system; microphone and amplifier public address system.

Tape recorders.

Radio.

Closed-circuit television.

The pattern of communications should be derived from a study of the systems available, the administrative structure of the undertaking, and the activities which are to be provided for. Such a study is usually a joint exercise with specialists, and may often reveal factors either not previously appreciated or ignored. Weaknesses may be exposed in existing systems, the adverse effects of which had not even been suspected in the operation of the undertaking.

Within industry, good communications can make a substantial contribution to safety, productivity and efficiency, and not least in the maintenance services. Speedy location of key personnel, the rapid transmission of instructions between supervisory and operational personnel, quick information relating to delays and stoppages, and faster remedial action, are among the more obvious gains.

*Chapter 8*

# COSTING AND FINANCE
## for
## Maintenance Control and Accountability

## INTRODUCTION

There are two spheres of activity which together constitute the financial operation and control of an undertaking:

> Cost accounting, which portrays in monetary values the activities of the undertaking; and Finance accounting, which is more directly concerned with the monetary transactions of the undertaking.

Earlier systems of financial accounting showed the position in terms of the profit and loss in a business, and were limited to statements of monetary transactions and stock-taking. They gave no information about the location or nature of the contributory factors bringing about the results presented.

Modern industrial organisations require a wider understanding of the principles, value and purpose of cost accounting in those who bear administrative responsibility. The science of cost accounting has become highly developed and is recognised as an important element of management.

This does not mean that engineers are expected to act as cost accountants, but in the same way that training and experience enables them to assess and criticise designs or drawings, so should they cultivate the ability to assess and interpret returns prepared by accountants which describe in financial terms the operations and activities for which they are responsible. These should be used as clues to the causes of waste and inefficiency, and to the ways in which improved performance and reduced costs can be obtained. They should be recognised as a means of regulating and controlling expenditure.

## BASIC TERMINOLOGY

### *Capital*

Wealth employed to produce more wealth; its value lies in its productiveness. It is variously described, according to how it is used. It may be augmented out of the proceeds of production to increase or improve the efficiency of future production.

### *Revenue*

The income derived from the employment of capital, and that portion of income which is expended in sustaining its continuous derivation from the capital.

### *Fixed assets*

These comprise such things as buildings, plant, machinery and fixtures. They are depreciable.

### *Liquid assets*

These are represented in such things as stores, spares and component holdings. They should be depreciated by revaluation at specified intervals. Apart from deterioration, there is the additional risk of loss or theft.

### *Depreciation*

The reduction in value of the fixed assets due to the lapse of time and unavoidable deterioration, wear and tear. It is a charge against current production.

### *Obsolescence*

This arises as the plant or equipment becomes out-of-date in terms of inadequate production, high operational cost, or uneconomic servicing.

### *Maintenance*

The upkeep of the means and facilities of production, including inspection, cleaning, lubrication, running repairs and adjustments, overhauls with new parts and components, and painting. It aims to keep production 'downtime' to a minimum and to achieve the highest economic level of efficiency from plant and equipment, with satisfactory utilisation factors. It prolongs the life of equipment, allows more favourable terms of depreciation, and defers the need for outlay on replacement of fixed assets.

## The elements of cost

### *Direct or prime costs*

Costs allocated directly to a job, such as the costs of material and labour directly employed in the production or carrying out of the job:

Prime Cost = Direct Labour + Direct Material + Other Direct Charges or Expenses

These costs can be established with a high degree of accuracy.

### *Indirect or overhead expenses (oncosts)*

Costs which cannot be allocated directly to a particular job, but which must be absorbed by apportioning a due share to all jobs. They include materials and labour used or involved indirectly, common services and facilities, establishment charges including administration; they are sometimes referred to as 'burden'.

### *Standard costs*

This is the term applied to pre-determined or budgeted estimates of cost based on existing conditions. They apply particularly to repetitive work, but are not so readily applicable to general maintenance work. They provide a means of assessing and controlling the efficiency of operations by highlighting where actual costs deviate from the pre-determined standards. From this aspect, their application should be considered wherever it would be beneficial; where they are applied, their periodic review (to take account of changed conditions) is essential.

### *Service departments*

This term denotes those sections of an undertaking which provide a service or utility to the productive sections. Examples include power stations, boiler plants, compressed air stations and central heating plants; workshops, stores, transport, canteens and medical centres.

## THE PURPOSES AND FUNCTIONAL FRAMEWORK OF A FINANCE DEPARTMENT

The establishment and procedures of finance departments vary widely, even in undertakings of similar types and organisation, but in general they have two basic purposes.

The first is to provide a means of cost accounting appropriate to the activities within the units and departments of the undertaking, and to furnish the essential data to those in charge of the administration of those sections, to enable them to control the financial aspects of their operations. The second is to establish the financial aspects of the undertaking as a whole, and to provide the data which the management needs to assess and control the financial aspects and operation of the business.

Figure 8.1 shows a typical system designed to meet these needs, and indicates the factors affecting the financial operations and control of an undertaking. There is of necessity a close link between the costing system and the financial accounts, as the two aspects must be reconciled.

## MAINTENANCE COSTING AND FINANCE

Since maintenance is a service to production, as distinct from production itself, it constitutes a production overhead in financial terms. Its efficiency is reflected in its effect on production generally. Similarly the importance of maintenance costs lies in their influence on production costs, and their presentation should be in production terms. This does not preclude the introduction of specific costing methods, or the initiation of particular cost exercises to explore and control the costs of maintenance. The costing system, and the data obtained from it, should establish the total costs of maintenance activities, simplify their control, and ensure their equitable distribution between the various sections of production activity.

## THE CLASSIFICATION OF MAINTENANCE COSTS

There is some divergence between the needs of the engineering establishment in controlling and accounting for maintenance costs, and those of the finance and production departments in incorporating them into their returns. The maintenance establishment is inevitably

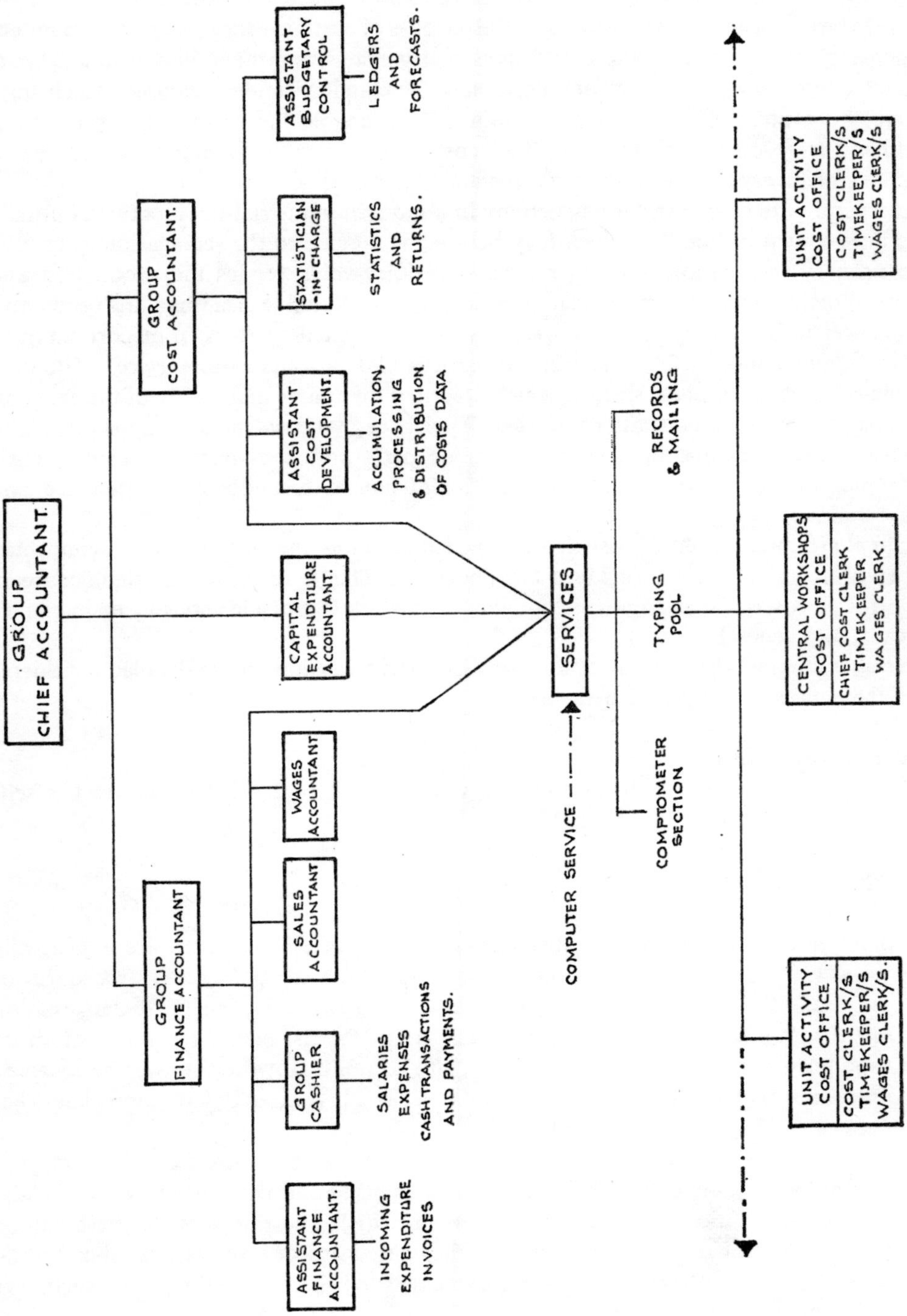


FIG. 8.1. Functional framework of a typical finance department.

involved in activities which are not definable as true maintenance. If the real costs of maintenance are to be established (which should be the purpose of costing) every category of maintenance department activity needs to be recognised, segregated and taken into account. From the engineering standpoint, therefore, costs must be classified under headings which both facilitate the control of engineering expenditure and enable the finance and production departments to recognise and interpret these costs, together with the activities from which they arise, for incorporation in their returns and statements.

The general pattern of engineering activity in a modern industrial undertaking is usually on the lines shown in Fig. 8.2, which may be used as a guide to the segregation of costs.

The primary segregation is by 'work classification' which relates the production and engineering activities at the particular production unit. Suitable headings are productive maintenance, utility maintenance, workshops and associated services, transport services, buildings maintenance, safety and welfare services and miscellaneous services. The next grouping is that of costs classification, which describes the source and nature of the expenditure under the various classifications of work. Typical capital costs include additional plant, renewals or replacements, improvements, major parts and components. Revenue costs include planned maintenance, unscheduled maintenance, breakdown maintenance and contract maintenance.

The final grouping is that of cost elements which describes the elements which contribute to expenditure under the various cost classifications. This is of particular significance in the control of costs; the elements include labour, materials, outside services, maintenance overheads, and general overheads.

The determination, distribution and control of costs necessitate more detailed consideration of these classifications and elements.

## Work classification

The items included under each heading are shown in Fig. 8.2. This breakdown is self-explanatory.

## Costs classification

### *Capital expenditure*

The differentiation between what constitutes capital or revenue expenditure is generally set out in a 'Capital and Revenue Code', conforming with the policy and procedures of the undertaking. While the grouping is quite clear-cut in many cases, the line of demarcation is sometimes rather obscure. The interpretation applied by the inland revenue authorities for taxation purposes is a factor which is taken into account. However, it must be accepted that maintenance includes some capital expenditure, and it is necessary to appreciate what this category includes.

The costs can be shown on the maintenance cost returns and statements. They may be interpreted differently for purposes of financial accounting and control, but it is essential that the basic information and accountability is provided. The provision of the building, structure, plant unit or major component, whether purchased or built, is treated as capital expenditure involving alterations to the fixed assets of the undertaking. As such, the appropriate depreciation rates are applied and levied as production overheads.

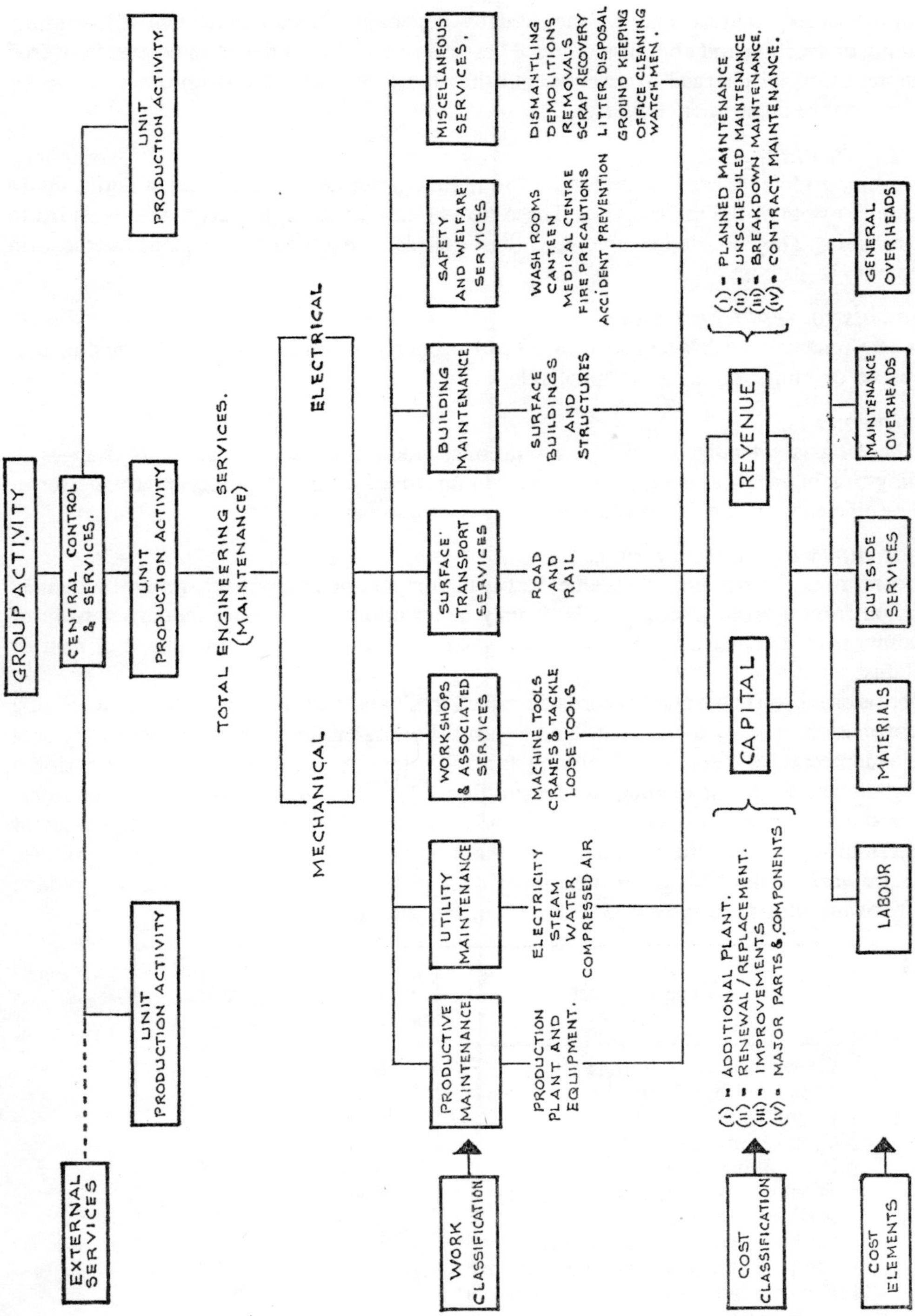


FIG. 8.2. Typical chart of engineering activity at a production unit.

Additional expenditure may be incurred by the maintenance department in handling, installing or erecting and commissioning. These may be added to the main cost and covered by depreciation, or charged to revenue, but they must be segregated from other categories of maintenance department activity.

### *Additional plant*

This heading includes costs incurred by the maintenance department due to additions to the capital assets of the undertaking. These may be new units of property, or extensions to existing units. They are depreciable, and these charges are added to the plant overhead in proportion to its cost.

### *Renewals or replacements*

Complete renewal or replacement of a unit of property is defined as capital, whether it is purchased or built. Again, it is depreciable.

### *Improvements*

This heading is intended to include alterations, modernisations or structural changes to buildings, plant and equipment, which result in improved durability, productivity, performance or efficiency. It may involve revised depreciation charges.

### *Major parts and components*

This comprises the straight replacement of a major part or component, generally defined as capital from a given level upwards. It may be an improved or more modern version of the component it replaces, and requires revised depreciation charges by reason of cost or durability.

There is provision in the plant records described in Chapter 2 for the essential data relating to depreciation charges to be recorded. Expenditure under these heads is frequently met from a depreciation reserve and sinking fund, accumulated out of revenue by periodic regular allocations, to make good the diminishing value of the fixed assets due to deterioration or obsolescence. The depreciation is spread over the serviceable life of the asset at predetermined rates for the various types, these being referred to as amortisation tables. There are various methods of providing for depreciation, depending on company policy. The following illustrates typical rates based on the 'straight line' method:

| Type of Asset | Period over which cost is written off | Annual Rate (straight line) |
|---|---|---|
| General office buildings | 50 yrs | 2% |
| Industrial buildings | 40 yrs | $2\frac{1}{2}$% |
| Locomotives | 20 yrs | 5% |
| Generating plants | 20 yrs | 5% |
| Plant and machinery | 15 yrs | $6\frac{2}{3}$% |
| Workshops plant and machinery | 15 yrs | $6\frac{2}{3}$% |
| Railway wagons | 12 yrs | $8\frac{1}{3}$% |
| Canteens (heavy equipment) | 8 yrs | $12\frac{1}{2}$% |
| Medical centres (heavy equipment) | 8 yrs | $12\frac{1}{2}$% |
| Road vehicles—lorries, vans and cars | 5 yrs | 20% |

The charges are levied as an overhead of the upkeep of the asset. The depreciation period starts when the property is commissioned, and the plant is then said to be 'capitalised'. The taxation authorities allow tax rebate under the heading of wear and tear, based on the official depreciation rates, which are not necessarily those employed internally.

### *Revenue expenditure*

This is expenditure incurred in maintaining the fixed assets and property of the undertaking in safe and satisfactory condition, and sustaining its production or service capability. Four sub-classifications are desirable for the purposes of control and accountability, namely:

#### *Planned Maintenance*

This includes scheduled maintenance as described in the procedures for maintenance and repair (Chapter 5); that is:

I — Inspection.
C — Cleaning.
L — Lubrication (so far as this is included in the planned schedules, as distinct from the operational function).
RS — Routine Servicing.
SO — Site Overhaul.
P — Painting.

#### *Unscheduled maintenance*

This includes:

RR — Running Repairs and Adjustments.
AH — Miscellaneous Ad-hoc Repairs and Overhauls.

#### *Breakdown maintenance*

This is the remedial work after breakdown to restore production.

#### *Contract maintenance*

This relates to maintenance work contracted out under service contracts or other arrangements.

Costs under these headings constitute the true maintenance bill and are charged directly to the production activity expenses. They are segregated as shown to provide a means of comparison and control. They include:

Direct expenses incurred or prime costs arising from work at site, at the central workshops, or done by external firms.

Indirect or overhead costs such as wages, salaries, superannuation, depreciation and common services such as power, heat and light, drawing office, transport, communications and administration.

Breakdown maintenance is the aspect on which particular attention must be focussed. The cost of the remedial work may be large or small, but the effect on production is direct, and the consequential costs in terms of lost production and unproductive labour can be very serious. Planned maintenance seeks to minimise the incidence of breakdowns, and these occurrences are a direct measure of its effectiveness.

## Cost elements

Each of the foregoing classifications will include one or more of the cost elements already referred to, namely:

### *Labour*

The direct cost of the time chargeable to a specific job or activity.

### *Materials*

Plant, Equipment, and Supplies classified as:

| | Stores Main Classification (see Chap. 6) |
|---|---|
| General Stores and Sundries | 1 |
| Plant Sub-assemblies, Spare Parts and Components | 2 |
| Plant Units or Composite Assemblies | 3 |
| Others | – |

### *Outside services*

External labour and supervisory services engaged in addition to the resources of the activity concerned, the costs of which include the provider's overheads. Engineering service and hire of mobile plant, tools and equipment may be included. The services of the central workshops may well be covered in this way, in common with the services of external firms.

### *Maintenance overheads*

Those costs of the maintenance department which cannot be directly allocated to a specific job, but which must be absorbed by apportioning a share to all jobs and operations. Examples include materials and labour used indirectly, common services and facilities, and establishment charges including supervision and administration.

### *General overheads*

Those portions of the undertaking's operational overheads allocated to maintenance in the process of their proportionate distribution to all production units; for example, a proportion of the costs of the purchasing and stores service; the transport service; accident prevention and safety; training; canteen services; medical services; testing and research; the secretariat; administration; and the engineering establishment.

In each case, these are operational expenses of the undertaking which cannot be allocated to a specific department or activity, but contain an element of cost to the maintenance services. In differentiating between maintenance overheads and general overheads, much depends on the nature of the undertaking, its organisation, and directive policy. A study of the functional organisation is necessary to ensure an equitable allocation of maintenance costs under these headings.

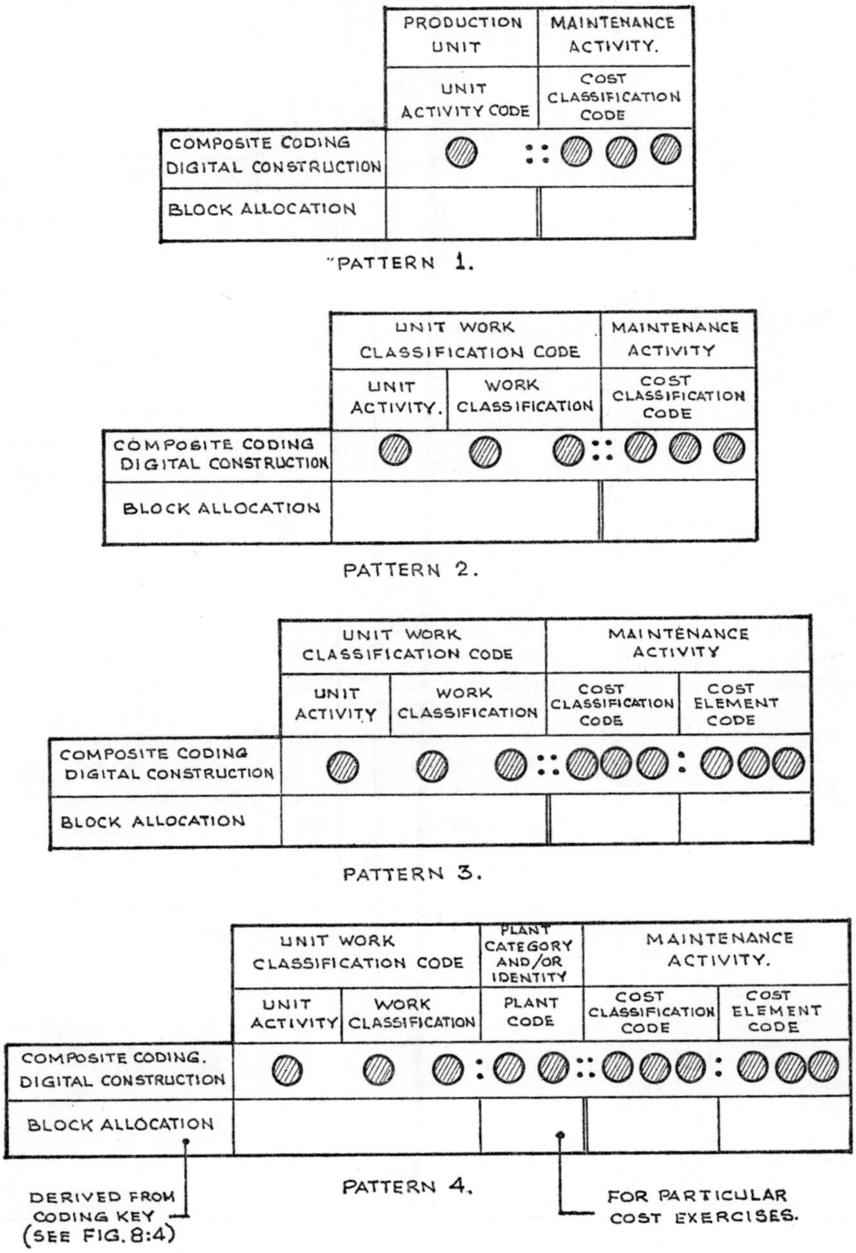


FIG. 8.3. Basic forms of composite coding.

**UNIT ACTIVITY**

| WORK CLASSIFICATION | CODE | |
|---|---|---|
| PRODUCTION: | 00 | |
| CUTTING AND LOADING | | 01 |
| ROOF SUPPORT | | 02 |
| FACE CONVEYING | | 03 |
| GATE BELT CONVEYING | | 04 |
| TRUNK BELT CONVEYING | | 05 |
| UNDERGRD. ROPE HAULAGE | | 06 |
| UNDERGRD. RAIL TRANSPORT | | 07 |
| LOADING STATIONS | | 08 |
| CREEPERS | | 09 |
| VENTILATION | | 10 |
| PUMPING | | 11 |
| DUST SUPPRESSION | | 12 |
| WINDING. | | 13 |
| SHAFTS | | 14 |
| DECKING UNDERGROUND | | 15 |
| SURFACE HEAPSTEAD. | | 16 |
| PREPARATION | | 17 |
| OUT-LOADING | | 18 |
| DIRT-DISPOSAL | | 19 |
| UTILITIES: | 20 | |
| ELECTRICITY | | 21 |
| STEAM | | 22 |
| WATER | | 23 |
| COMPRESSED AIR. | | 24 |
| WORKSHOPS: | 30 | |
| MACHINE TOOLS | | 31 |
| CRANES & LIFTING TACKLE | | 32 |
| LOOSE TOOLS. | | 33 |
| SURFACE TRANSPORT: | 40 | |
| ROAD | | 41 |
| RAIL. | | 42 |
| BUILDINGS: | 50 | |
| INDUSTRIAL | | 51 |
| NON-INDUSTRIAL | | 52 |
| SAFETY & WELFARE: | 60 | |
| BATHS AND WASHROOMS | | 61 |
| CANTEEN. | | 62 |
| MEDICAL CENTRE | | 63 |
| FIRE PRECAUTIONS | | 64 |
| ACCIDENT PREVENTION | | 65 |
| MISCELLANEOUS: | 70 | |
| DISMANTLING, DEMOLITION, AND REMOVALS | | 71 |
| SCRAP RECOVERY. | | 72 |
| LITTER DISPOSAL | | 73 |
| GROUND-KEEPING | | 74 |
| OFFICE CLEANING. | | 75 |
| WATCHMEN. | | 76 |

ADD INITIAL DIGIT IN EACH CASE TO IDENTIFY PRODUCTION ACTIVITY.

**MAINTENANCE ACTIVITY**

| COST CLASSIFICATION | CODE | | |
|---|---|---|---|
| CAPITAL: | 200 | | |
| (I) ADDED BUILDING, STRUCTURE OR PLANT. | | 210 | |
| (II) RENEWAL/REPLACET. OF BUILDING, STRUCTURE OR PLANT. | | 220 | |
| (III) IMPROVEMENTS. | | 230 | |
| (IV) MAJOR PART OR COMPONENT. | | 240 | |
| REVENUE: | 300 | | |
| MAINTENANCE: | | | |
| (I) PLANNED: | | 310 | |
| INSPECTION | | | 311 |
| CLEANING | | | 312 |
| LUBRICATION | | | 313 |
| ROUTINE SERVICING | | | 314 |
| SITE OVERHAUL | | | 315 |
| CYCLIC REPLACEMENT | | | 316 |
| (II) UNSCHEDULED: | | 320 | |
| RUNNING REPAIRS & ADJUSTMENTS. | | | 321 |
| AD HOC REPAIRS & OVERHAULS. | | | 322 |
| (III) BREAKDOWN: | | 330 | |
| (IV). CONTRACT: | | 340 | |

| COST ELEMENTS | STORES CLASS | CODE | | |
|---|---|---|---|---|
| LABOUR: | | 400 | | |
| S. SKILLED | | | 410 | |
| SS. SEMI-SKILLED | | | 420 | |
| US. UNSKILLED | | | 430 | |
| MATERIALS: | | 500 | | |
| (I) GENERAL STORES & SUNDRIES | 1. | | 510 | |
| (II) SUB-ASSEMBLIES, SPARE PARTS & COMPONENTS | 2. | | 520 | |
| (III) PLANT UNITS OR COMPOSITE ASSEMBLIES. | 3. | | 530 | |
| (IV) OTHERS | 4. | | 540 | |
| OUTSIDE SERVICES: | | 600 | | |
| CENTRAL WORKSHOPS: | | | 610 | |
| LABOUR | | | | 611 |
| MATERIALS | | | | 612 |
| ONCOSTS. | | | | 613 |
| OUTSIDE FIRM: | | | 620 | |
| LABOUR | | | | 621 |
| MATERIALS | | | | 622 |
| ONCOSTS | | | | 623 |
| PLANT HIRE | | | | 624 |
| MAINTENANCE OVERHEADS: | | 700 | | |
| INDIRECT LABOUR | | | 710 | |
| INDIRECT MATERIALS. | | | 720 | |
| SUPERVISION | | | 730 | |
| ADMINISTRATION | | | 740 | |
| GENERAL OVERHEADS: | | 800 | | |
| PURCHASING & STORES | | | 810 | |
| TRANSPORT SERVICES | | | 820 | |
| SAFETY & WELFARE DEPT. | | | 830 | |
| TESTING & RESEARCH | | | 840 | |
| SECRETARIAT | | | 850 | |
| ADMINISTRATION. | | | 860 | |
| ENGINEERING ESTABLISHMENT | | | 870 | |

FIG. 8.4. Typical cost coding key.

## THE CODING SYSTEM

To ease the determination, distribution, and control of maintenance costs, a suitable coding system should be established. This must be suited to the organisation of the undertaking as a whole, as it helps in relating maintenance activities and costs to those of production.

The production establishments of the undertaking, and the work classifications within them, must be coded in a manner that segregates them by units, operations, or plant. Similar items at different Units should have a common coding, modified to identify the particular Unit concerned. Further, these operations and activities must be allotted the appropriate cost account classifications already described, which must be so coded as to segregate them.

If these two requirements are met, costs can be equitably distributed and controlled with any desired degree of refinement. Purposeless analysis must, of course, be avoided.

The basic form of the composite coding to meet these requirements is shown in Fig. 8.3, the actual pattern being selected according to the degree of refinement required. The latter can be decided in both production and cost codings, so that the final composite coding provides for objective analysis and control.

Such a system lends itself to mechanised methods and the employment of computers. The selection will probably be made from the first three patterns, 2 and 3 offering a compromise between the coarseness of 1 and the refinement of 4. The latter could be applied in specific cases and costing exercises if desired. The block allocation of digits can best be decided by preparing a costs coding key for the production unit concerned. Figure 8.4 illustrates such a coding key applied to mining practice, but it can be adapted to suit other applications. An undertaking which includes numbers of similar production units can use the key throughout, the only variable being the initial digit of the code identifying the unit concerned. This common coding can be very convenient in large concerns.

The work classifications side of the coding key can be varied to meet other applications in the same industry or in other industries, but the remainder may well be retained in all cases.

## THE RECORDING, ACCUMULATION, PROCESSING, AND DISTRIBUTION OF COSTS DATA

### Plant, equipment and supplies

These come within the following stores classifications:

General Stores and Sundries.
Plant Sub-assemblies, Spare Parts and Components.
Plant Units or Composite Assemblies.

Invoices relating to purchased plant, equipment, and supplies bear an order number allocated by the purchasing department (see Chapter 6) which relates the account to an internal revenue requisition (Fig. 6.6) and an internal capital requisition (Fig. 6.7). On each of these is entered the appropriate charging instruction; for example:

### *For revenue items*
The stores main classification; the central workshops job number (for internal manufacture or processing against a works order); and the costing code.

### *For capital items*
As above, together with the project authorisation number, section, and item.

As the invoices from the various suppliers have different sizes and lay-outs, it is convenient to attach to each a small standard invoice check slip with gummed edge, on which the charging instruction can be entered from the internal revenue or capital requisition. This simplifies the processes of costs accumulation, processing, and distribution. A typical slip of this type is shown in Fig. 8.5. In the case of items held in stock accounts, the charge to maintenance does not arise until they are actually issued for use through the medium of the stores warrant, Fig. 6.12. This carries the appropriate charging instruction. When there is a centralised purchasing and stores service, with stores at the various Units, there is a sound case for treating all items as received into stock and withdrawn under a stores warrant.

In any case, all invoices and stores warrants are accumulated under the appropriate costs classifications; these are summarised, computed, and distributed monthly under the related Production Activity (see note at end of section on 'Outside Services' below).

## Labour
The direct cost of the labour employed in the various occupational categories of maintenance within the undertaking accrues from operations at the central workshops and the various production units. In Chapter 4, reference is made to the operative's weekly clock card (Fig. 4.25) and weekly time allocation card (Fig. 4.24). These must be reconciled in the checking process. The time is allocated against workshop job numbers, obtained from works orders (Fig. 4.12), which bear the appropriate charging instruction. The costs are accumulated against each works order and the summarised costs under each maintenance classification are computed and distributed monthly under the related production activity. The workshop documentation and procedures already described simplify the accumulation of the total job cost under the works order, and the labour element of cost is there shown as part of the total costs.

The total job cost is a measure of the workshop performance, whereas the processing and distribution of these costs under the various maintenance classifications relate them directly to production costs at the Units, and thereby facilitate control.

Similar weekly clock cards (Fig. 4.25) and weekly time allocation card (Fig. 4.24) for site work are described in Chapter 5; again, these must be reconciled weekly, as with the central workshops. The job numbers in this case are those used for work allocation and scheduling purposes. A charging number is entered against each job number and this serves to record, accumulate, and distribute the labour costs under the various maintenance classifications (see note at end of next section).

## Outside services
The forms which these can take are given in Chapter 5. The method of accumulating, processing, and distributing these costs can be similar to that described earlier in this Chapter for purchased plant, equipment, and supplies, the rendered invoice or account being the

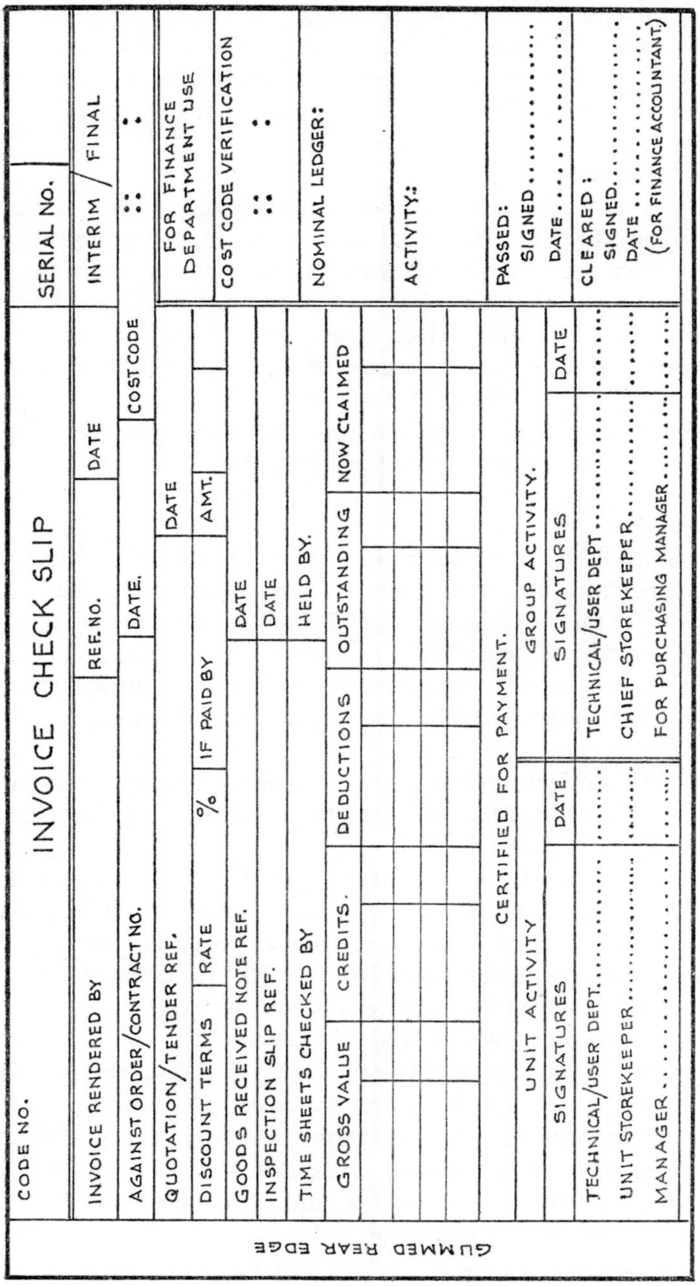


FIG. 8.5. Invoice check slip.

initiating document. It may be advisable to have these incoming accounts subdivided into materials, labour, and oncosts, with miscellaneous expenses such as travelling, lodging and subsistence shown separately.

NOTE: Referring back to the sections describing plant, equipment and supplies, labour, and central workshops service charges, an alternative is to treat the completed job cost as an external invoice, as described for outside services.

## Maintenance overheads

This comprises those costs of maintenance operation and practice which cannot be allocated directly to a specific job. They include such items as plant, equipment and supplies, indirect labour, salaries, superannuation and pensions, holiday pay, national insurance rates, depreciation, power, heat and light, and so on. These must be recorded, accumulated as maintenance overheads, and distributed on some equitable basis which is related to the circumstances applying and the accounting policies of the undertaking concerned. The methods adopted are varied; for the central workshops, the basis may be the direct man-hours expended. For the welfare facilities the basis may be the numbers of people served, while that for the heating and lighting installations may be the floor space.

## General overheads

The sources from which costs accrue under this heading are set out earlier in this Chapter. The maintenance department is not normally involved in the records and procedures for dealing with costs under this heading; it is simply allocated its share on an agreed basis.

From these data, periodic and cumulative analyses of the total maintenance expenditure may be prepared. The engineering costs summary in Fig. 8.6 illustrates such an analysis applied to a typical Unit. Such a return can be prepared monthly, quarterly, half-yearly, or yearly, as desired.

The total of the separate Unit returns gives the total maintenance costs for the Group, and furnishes the Group's headquarters with the means of comparison and of judging whether and where remedial action is called for.

The analysis shown provides for a fully detailed breakdown of costs, but the means are provided either for simplification or for fuller information as desired. For example, 'Production' may be broken down into its elements as coded in Fig. 8.4; 'Capital' and 'Revenue' may be accumulated under codes '200' and '300' respectively; and cost elements may be accumulated under labour (400), materials (500), and so on.

Provision is made for relating costs to production level.

The actual and budgeted costs may also be compared; this is discussed in the next section. It is likely that a general summary of maintenance costs will suffice for incorporation in production cost statements and returns. As regards the accountability of the maintenance management to the production and general management, the engineering establishment can trace the source of any expenditure queried, through the medium of the cost element analysis provided, and can give the necessary explanations. It is also possible to indicate those cases where the remedial action lies outside the control of the maintenance management, and the form that such action should take.

The documentation and procedures for the accumulation and distribution of maintenance

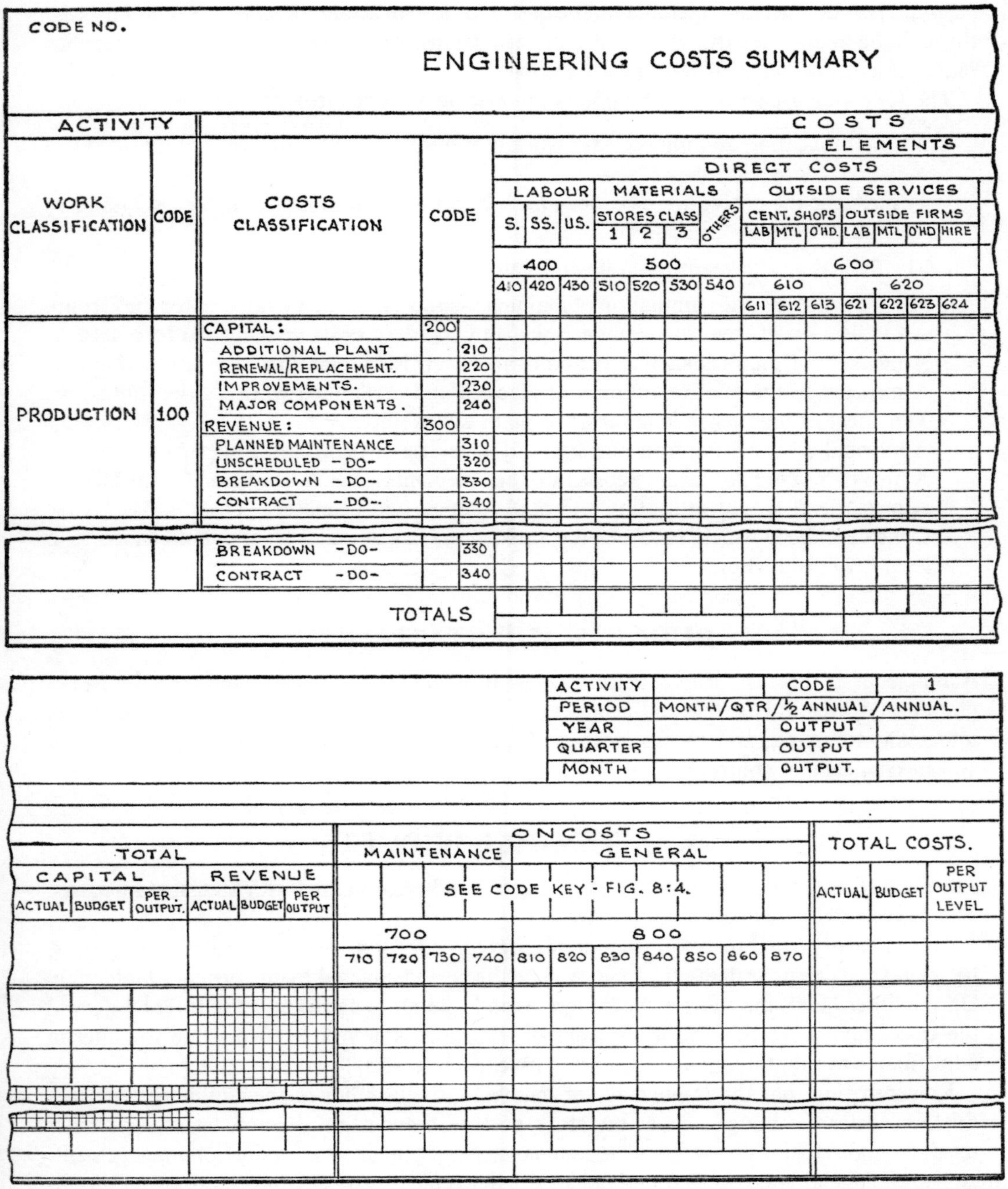


FIG. 8.6. Engineering costs summary (for breakdown of work classifications in first column see key to Fig. 8.4).

costs are best devised and operated on the basis that it is an accounting process, administered through the finance department, but established in close collaboration with the maintenance management, so that their joint needs are met.

The following are some guiding points in devising a procedure:

Avoid duplication and unnecessary work in meeting the differing needs of the accounting and maintenance department.

Ensure that the procedure provides timely information to both departments, in a form which aids effective control.

The data should present the following information:

A broad analysis and summation of maintenance costs for each Unit, and for the Group.
An indication of trends or comparisons with standard costs or budgeted forecasts.
A measure of the incidence of breakdowns in terms of their effect on production.
A measure of the relative volumes of planned and unplanned work carried out.
The effect of maintenance in terms of plant availability.
Abnormal costs arising from unsuitability or abuse of plant and equipment.
A breakdown of maintenance costs into their elements for each Unit, and for the Group.
A measure of labour utilisation and performance, and of material costs.
Control indices which promote the satisfactory accountability of maintenance as a service to production.

## BUDGETS AND STATISTICS

A budget is a forecast of expenditure. It is, in effect, the financial equivalent of works planning. Statistics are the end-product of the costing system, and illustrate both the information derived, and its significance. Together, they constitute tools of management, and mechanisms of control.

## MAINTENANCE BUDGETS

The main classifications, elements, and codings should be compared with budgeted forecasts. This is essential for effective cost control.

The financial procedures of most undertakings provide for a fiscal period of twelve months, between fixed calendar dates. This can well be the period covered by the primary budgeting. On the other hand, the shorter the budget period, the more accurate is the forecasting and the easier and more effective is the control. Hence it is advisable to provide for interim adjustments at monthly, quarterly, and half-yearly intervals.

Remembering that the importance of maintenance costs lies in their influence on production costs, the budgets must therefore portray them in production terms, and must result from discussions between the maintenance management, the production management, and the financial control. The maintenance management provides information relating to expenditure under the various maintenance classifications. The production management gives information relating to production targets, and plant, equipment, and service requirements. The finance control completes and correlates the information, prepares the budget statements, and issues the periodic returns and statistics relating to actual performance against forecast performance.

When the central workshops are operated on the lines discussed in Chapter 4, and site maintenance is organised and operated as discussed in Chapter 5, the preparation of the budgets is relatively simple. However, past and present performance and experience must be modified to allow for such variables as likely variations in labour rates, and prices of materials; adjustments in production schedules, in terms of output, load or demand on existing plant, level of utilisation, and reduction or increase in plant requirements; and improvements or modernisation of plant, methods and processes.

## MAINTENANCE STATISTICS

These should present as simply as possible, the information with which effective control of operations and costs can be established. They may be presented in diverse forms according to their nature and purpose. They may display results in tabulated form, but are often more effectively presented pictorially as charts or graphs. Trends and relationships stand out more clearly in the latter.

The significance and interpretation of the costs data described in the foregoing sections are further discussed in Chapter 9.

## STAFFING

The scale of staffing and the allocation of duties to cover the costing procedures will depend upon the nature and size of the undertaking, its general organisation, and the directive policies in the field of financial control. The degree of mechanised accounting introduced will, of course, be an influencing factor. The establishment may be based on the functional diagram shown in Fig. 8.1, tailored to suit the circumstances applying. The disposition of cost offices, control centres, and personnel is established to suit the documentation and procedures introduced.

## OPERATIONAL PROCEDURE AND CONTROL

These are conveniently illustrated by a series of charts covering the elements of cost contributing to the total. For clarity, these can be prepared under the following separate headings:

CHART 1 — Labour (excluding that under outside services submitted by rendered invoice and works orders returned by the central workshops).

CHART 2 — Plant, equipment, and supplies (stores classifications 1, 2, and 3, and outside services).

CHART 3 — Overheads, costs summaries, statistical and budgetary statements.

From these, the final form and flow of documentation, personnel, and duties can be established.

*Chapter 9*

# GENERAL ADMINISTRATION with Guides to Efficiency

## ESTABLISHMENT OF AUTHORITY

Overall administrative control usually rests with the engineering establishment, with its chief answering to production management. This responsibility may be delegated within the engineering establishment; for example, maintenance (as distinct from projects and developments) may be the departmental responsibility of a senior engineer.

## QUALITY OF LEADERSHIP AND SUPERVISION

The organisation, procedures, and practices instituted to regulate the maintenance activities and demands in an industrial undertaking are not in themselves a guarantee of satisfactory results. The senior executive and his staff must influence the whole functional activity. Maintenance performance can never rise above the quality of its leadership and supervision. From good leadership stems the team-work which is the essence of success in any such enterprise. Talent and ability must be recognised and fostered; good work must be noticed and commended; and shoddiness, neglect, and carelessness must be exposed and dealt with.

## GRAPHICAL AIDS TO ORGANISATION AND CONTROL

In organisation, establishment, and control, graphical aids can be of great value. They have an immediate appeal to engineers, and can be designed to make facts clearer, to enable a grasp of a situation or plan to be obtained rapidly, and to simplify scheduling of work, deployment of manpower, and control.

## THE CULTIVATION OF THE COST SENSE

The cultivation of cost-consciousness at all levels of supervision through an understanding of the financial results of their actions and operations, from data suitably presented, is of primary importance.

## THE SELF-CRITICAL APPROACH

A healthy departmental self-critical approach to the effectiveness and efficiency of its contribution and performance is to be encouraged, by considering such questions as:

What was it aimed to do?
How far is this being, or has been, achieved?
Where has the shortfall occurred, and why?

Method study and operational research can be useful for this purpose.

## THE STANDARD OF MORALE

Morale is one of those intangible evidences of discipline and spirit which is sensed in moving among, and talking to, the personnel in an undertaking. Good morale springs from true leadership, sound administration, and known purpose, direction, and achievement. The humanising of contacts, often seen to be so effective in smaller undertakings, may give way to feelings of isolation, frustration, and bureaucratic relationships in larger concerns. The aim should be to generate a sense of mutual concern and interest in human relationships.

## INVESTMENT POLICY

An adage of the world of finance is that 'Money spent wisely makes money'; a sound investment policy translates this into reality. Technical developments, new machines, new materials and methods, can be used to reduce the cost of maintaining production facilities, to improve performance, and to raise safety standards. The best machinery is likely to be the most economical in the long-run. In making a case for this, it is sometimes more effectively presented in terms of buying output, rather than merely purchasing plant and equipment.

## INCENTIVES

The varied nature of the maintenance tasks, and the differing needs and conditions arising, together with the influences of production activity, are not attuned to the adoption of incentive systems of payment. There are, however, some directions in which their application can be usefully considered. One obvious case is that of repetitive work in the central workshops. The forward planning of maintenance work can sometimes lead to an incentive payment arrangement, based on the completion of known tasks in a given period, but care must be taken to ensure that the required standards of work are not endangered. In some cases, maintenance incentives can be included in output bonus schemes, by arranging that continuity of production, and attainment of targets, provides rewards to both production and maintenance personnel.

## BUDGETS

Budgeting is basic to cost control. It forms a basis for the judgement of actual performance, and through cost control it shows if remedial measures are necessary. The real costs of maintenance are not easily assessed. Its true costs should be segregated from those of the indirect activities of the department, if satisfactory control and accountability are to be established (see Chapter 8).

## STATISTICS

If statistics are to be used effectively, they must be accurate and readily intelligible. Their usefulness depends on their interpretation. The data used to translate operational activity into terms of productivity and profitability have three important facets; namely, collection of data, presentation, and interpretation. To be useful, the facts must be right, accurately presented, and correctly interpreted both for what they ARE and for what they are NOT.

## SAFETY

Safety goes hand-in-hand with efficiency. An efficient method is a safe method. Apart from the paramount consideration of safety of personnel, accidents cause consequential losses in terms of compensation, lost output, and damage to plant and equipment. Safety is a matter for constant emphasis and vigilance.

## PERSONNEL ADMINISTRATION

This is illustrated in Fig. 9.1. Major factors in its successful application include the selection and deployment of the right men at all levels; conditioning the working environment so that full scope is given to skill, initiative, and enterprise, and developing the qualities of leadership and ambition.

Craftsmanship is highly important in all industrial activity. It is incumbent upon professional engineers and technicians to foster the co-operation of the craftsmen under their control, through mutual respect, understanding, and appreciation of the important role of each.

### Selection of personnel

The changing techniques of industry, and its complexity, increase the need for selectivity in meeting the personnel requirements at all levels. Furthermore, the careful selection of personnel for their vocational fitness, medical fitness, and personal attributes, contributes towards satisfactory human relationships within an undertaking; these in turn have a marked influence on efficiency.

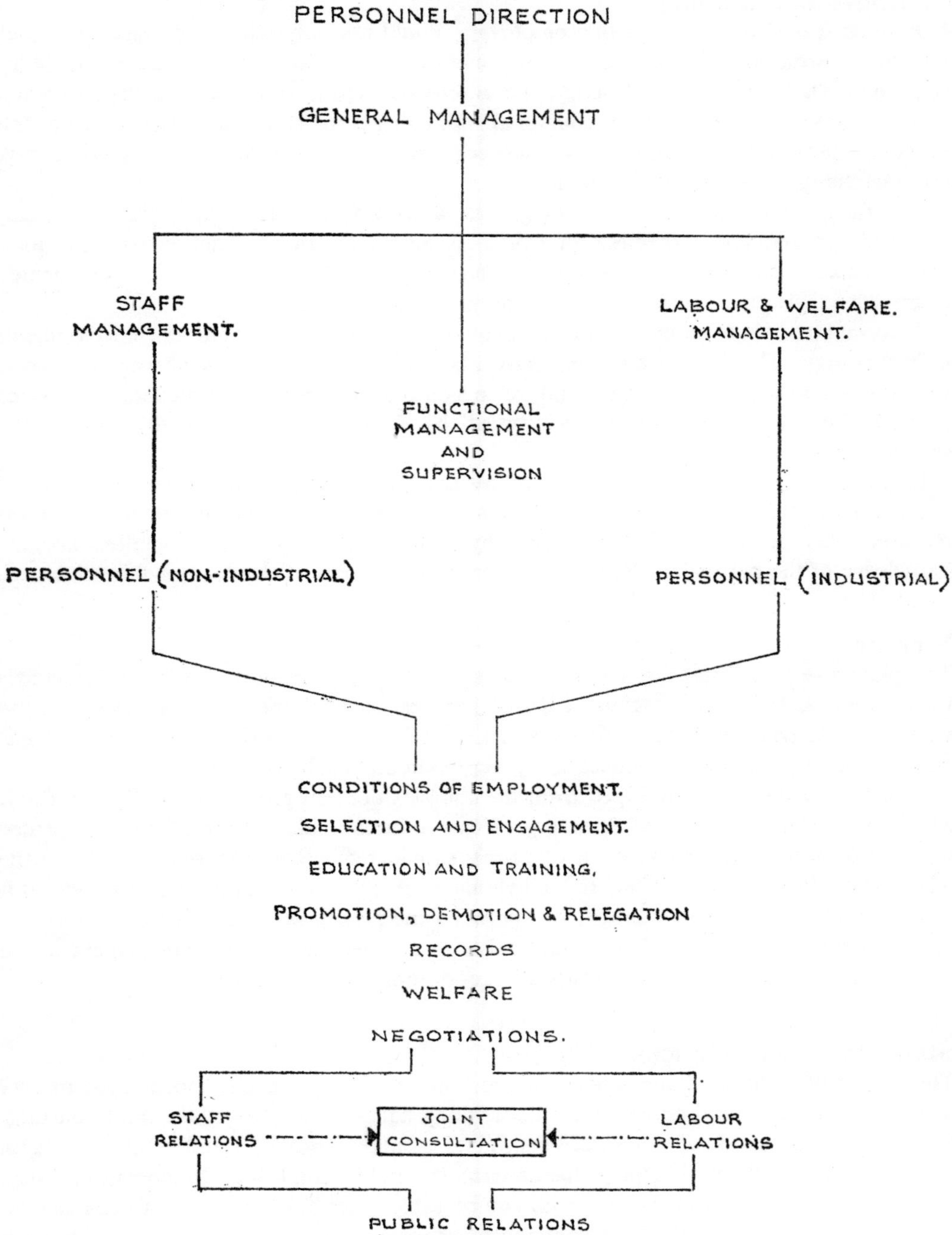


FIG. 9.1. The field of personnel administration.

## Education and training

Nowadays it is also recognised that employers should not only select and place personnel, but should promote schemes and provide facilities for their further education and training, so as to increase individual proficiency, and to provide recruits for the supervisory and senior grades. For senior staff, refresher courses at staff colleges or schools are often a part of this process. The courses comprise lectures on specific aspects of their work; they also encourage the interchange of ideas, and discussion.

The further education of technical grades, craft workers, and apprentices is usually achieved through joint schemes, sponsored by employers in conjunction with the local education authority. Employees should be encouraged to take advantage of these schemes, to improve their proficiency and promotion prospects.

A normal trade background is often inadequate to cope with the continuing developments in technology. The increasing complexity and importance of maintenance engineering warrants a marked increase in training of machine operators and maintenance craftsmen through formal school courses, reinforced by informed instruction by experienced supervisors.

Efficient and economic production requires that plant and equipment must be operated and maintained correctly. When breakdowns occur, rapid diagnosis and remedial action is required; this challenge can be met only by a full understanding and detailed working knowledge of the particular plant or equipment, on the part of the personnel involved.

## Welfare

This plays a very important part in the achievement of good human relationships. It includes such aspects as safety and hygiene, catering services, medical and dental services, pension schemes, industrial thrift plans, financial assistance schemes, housing, personal problems, library services and internal publications, recreation and social activities.

Industrial welfare has developed from the complexities and growth of modern industry, which brought in their train increasing problems in human relationships. Those concerned with the direction and management of people should remember that welfare starts in the office, or on the shop floor, where the individual works; it is in his daily environment that it has its foundations. This establishes a link between line-management and the welfare services, and the correct conception that welfare exists to augment and assist management in the establishment of healthy environments and good human relationships.

## Staff and labour relations

The success of an undertaking depends significantly on the care taken to form a community of well-informed, keen, and lively people working harmoniously together. Joint consultation, in an atmosphere of free discussion, gives mental satisfaction and creates the team spirit necessary to the well-being of the undertaking and its employees. In modern industry, joint consultation has been applied with considerable success, in the form of management conferences, works councils, joint conferences of management and workpeople, and other groupings. Top-level management should be closely identified with these activities, if they are fully to achieve their purpose.

### Public relations

Relations with public bodies, the press, and government departments, are another aspect of management, often administered as far as personnel matters are concerned through the staff or labour department. For other considerations the secretariat may be the responsible department.

## MAINTENANCE EFFICIENCY

Despite all the organisation, planning, and systems of control, in the final analysis the standard of maintenance and its achievements rest mainly in the hands of the personnel and their chargehands or immediate supervisors. It is in the machine or plant itself that everything culminates; all else is a means to an end, namely the economic sustaining of the facilities of production and its ancillary services, with high safety standards. The organisation times the operations, and co-ordinates the functions involved, but the operative carries out the tasks. It is desirable that in doing so, he should know that his training, skill, and experience is being used. Hence the operations schedules will indicate the operations and their timing, but not the details of how they are to be carried out. There should be selective deployment of skilled, semi-skilled, and unskilled labour. It is not uncommon to find trained craftsmen performing such menial tasks as lubrication, cleaning of dismantled parts, moving of plant and equipment, clearing up after overhauls, and so on. The segregation of tasks into occupational categories (Chapter 5) is designed to overcome this undesirable state of affairs, and to make full use of the skilled labour available.

With managerial and supervisory staff, their time and energy should not be so much employed in coping with menial administrative tasks that they lose contact with the working force over which they should exercise a close and continuous influence. A periodic check on the time they are devoting to clerical work and menial tasks that could be carried out by more junior staff is recommended.

Evaluating maintenance performance is not easy. It is important to ensure that the work being done is in fact necessary. Schedules and programmes must be justified and based on practical experience in the conditions applying.

## VISUAL AND PHYSICAL GUIDES

To experienced eyes and ears, an indication of the physical standard of maintenance is gained at once from a tour of the buildings, plant, and equipment involved. The many indications include cleanliness, tidiness, the 'beat' of machinery, performance standards as registered by instrumentation and recorded in log books, visual evidence of attention to safety, such pointers to carelessness and neglect as rust and corrosion, loose bolts, nuts with insufficient hold, oil leaks and spillage, waste of salvageable surplus material, refuse dumped from so-called finished jobs, and evidences of inefficiency in the scrap bins and dumps of discarded or idle equipment awaiting either a decision or reconditioning and redeployment.

## ECONOMIC GUIDES

Much useful information can be derived from the engineering costs summary shown in Fig. 8.6. This includes:

The distribution of maintenance effort and resources between production proper and other work; capital and revenue work; and the various categories of the latter.
The utilisation and distribution of labour resources in terms of skilled, semi-skilled, and unskilled personnel.
The sources of demand and scale of use in the various classifications of stores.
The scale of outside resources employed, in terms of the locations, whether central workshops services, or those of outside firms are used, and their cost make-up.
The direct costs, oncosts, and total costs, for the whole of the maintenance activity.
The comparison of actual expenditure and budgeted forecasts.

Other information may have to be used and expressed in a different way, for presentation, interpretation, and control. This may include:

The planned effort in relation to that scheduled.
The incidence of breakdowns; causes, maintenance effort applied, nature, time, and cost of remedial work, and consequential losses.
The analysis of excessive material costs.
The variation of individual and total costs from budget forecasts, and the explanation.
The portrayal of cumulative trends, variations, and deviations, to give timely guidance as to where remedial action should be applied.
The labour cost analysis in terms of normal hours and overtime, and where applied.

The following returns relating to the central workshops performance and staffing can provide useful information:

Monthly return of performance and position at close of month, in terms of department, section, plant category, dispatches made, dispatches ready, work in hand, and work awaiting attention, all in terms of planned periodic overhauls and other activities.
Monthly return of works orders position at close of month, in terms of the numbers of orders received and completed during the month, and the number outstanding at the close of the month.
Monthly return of overtime worked in man-hours in the various departments and sections of the shops.
Monthly return of personnel in terms of shop sections, occupations, number on books at end of the month, numbers leaving and starting during the month, and vacancies arising from wastage and additions.
Cumulative annual throughput summary, month by month, for all types of plant.

These demand collaboration between management, engineering, and cost accounting, to evolve maintenance efficiency indices suited to their needs, and through them, to establish satisfactory control and accountability.

*Appendix I*

# DIRECTIVE OF PROCEDURE
## (For General Guidance)

| Designation of Post | Location |
|---|---|
| Workshops Manager | Group Central Workshops |

1. Receive monthly advice notes from Group engineering office advising machines due in and replacements due out to release them, and take appropriate action.
2. Receive and scrutinise all incoming works orders and drawings. If necessary, refer back to appropriate Group engineering office. If satisfied, initial, date, and pass to planning section for action with any additional instructions deemed necessary.
3. Examine and approve all outside purchase requisitions submitted by planning, and pass to appropriate Group engineering office in duplicate. Issue copy orders when received, to planning.
4. Examine and countersign all material urging notes submitted by progress and pass to purchasing department in duplicate.
5. Examine and countersign all transport advice notes submitted by progress, and pass to transport officer in duplicate.
6. Examine completed job records submitted by progress as to delivery, costs, etc., taking any appropriate matters up. Mark those for retention and route to records store via Group plant records officer. Destroy remainder after ...... month/s.
7. Supervise and direct all workshops administrative staff, and shop foremen, and maintain internal liaison.
8. Maintain liaison with Group engineering office, transport officer, purchasing and stores, and finance.
9. Maintain liaison with staff department, labour department, union representatives, and welfare matters generally.
10. Arrange and act as chairman of the interviewing panels for new workshops personnel.
11. Sit as chairman on consultative committee, take necessary action, and circulate minutes.

12. Submit monthly statistics and report to Group chief engineer in the form required by him.

13. Initiate or consider suggestions for any improvements in methods, materials, design, or procedure, and discuss with Group chief engineer if appropriate.

Authorised by ...............................................
*Group Chief Engineer*

Date of Issue .................

## NOTE

Directives of procedure, on similar lines, to be drawn up and issued for all central workshops staff in management and supervisory category.

*Appendix II*

# NETWORK TECHNIQUES CRITICAL PATH ANALYSIS (CPA)

Critical path analysis (CPA) is a modern network technique originating in the USA for the planning, co-ordination, and control of large and complex projects involving very considerable capital expenditure. In recent years, it has been widely used in Britain. The construction industries have led the way here in its application to large and one-off projects which present problems in planning and control.

In the USA, the system was developed in connection with two very different undertakings; namely the control of a missile development programme and the planning of the shut-down and reconditioning of a chemical plant.

The latter leads to a consideration of its possibilities for maintenance planning. There are two major differences between planning for project and development work and the planning of maintenance operations. First, in maintenance, the time element is concentrated so that there is little opportunity to adjust the plan, and exploit the facility to do this, offered by this network technique. On the other hand, maintenance is to a large extent repetitive, and there is time between each occasion to adjust and improve the plan. Hence it can well be applied to such operations as the annual shut-down of production plants for overhaul, the bi-annual overhaul of a steam turbine, the repetitive reconditioning of plant units at the central workshops under the periodic overhaul programme, or the site planned maintenance schedules. In each case the aim is to establish the most efficient arrangement of the operations in the job, so as to complete it as quickly and economically as possible, with the most useful deployment of manpower and facilities. The network can be quickly changed, with the aid of a computer, to accommodate unforeseen contingencies. This is claimed to be one of the main advantages of critical path analysis.

Network diagrams for major jobs, involving many operations, can appear to the uninitiated to be very complex, and the principles employed should be applied to a relatively simple job with a limited number of operations, before embarking on more complicated applications.

The following conditions are primary requirements:

The job must have a single start and finish.
The various operations in the job must be established.
The period involved in each operation must be assessed.
The sequence of the operations must be determined.

Hence a practical knowledge of the processes and procedures involved in the job, and experience of the time factors to be allowed, are necessary.

A network is a diagram which illustrates the inter-relation of the operations comprising a job. It differs from the conventional bar chart, in that it is not usually set down to a time scale, the lines indicating the relationship of operations, and not their duration.

It is not proposed to develop the subject further here. The reader is referred to the various publications now available on the subject.

*Appendix III*

# SERVICE MAINTENANCE AGREEMENT

| Contractor's Name and Address | Client's Name and Address |
|---|---|
| Telephone No. ........... | Telephone No. ........... |

Contract No. ............ Type of Contract ..................................................

Plant or Equipment Covered: ..................................................................

..........................................................................................................

..........................................................................................................

---

AN AGREEMENT made between ............................ Ltd. in the County of ...........

................ (hereinafter called 'the Company') of the one part and ............................

..........................................................................................................

(hereinafter called 'the Client/s') of the other part WITNESSETH as follows

1. In consideration of the payments hereinafter agreed to be made by the Client/s, the Company will from noon on the ................ day of ................ 19... until this Agreement is determined as hereinafter provided

   (a) ........... times during each year at approximately ........... intervals during the Company's normal working hours inspect ............................ as described in the Schedule and after each of such inspections report to the Client/s upon the condition thereof and upon defects therein apparent (if any) on external examination.

(b) In addition to the inspections aforesaid make during the Company's normal working hours such further inspections and reports on the said plant as the Client/s shall reasonably request.

(c) Provide a competent mechanic to make such inspections who will oil clean and adjust the mechanism of the said plant and carry out such small adjustments as may be effected by the mechanic using his normal hand tools and without the use of new materials.

(d) Provide without charge such grease and oil (excluding gear oil) and fuse wire as may be required for operating the plant.

2. If due to any strikes or industrial dispute competent mechanics are not available and the Company is unable to carry out the inspections detailed in Clause 1 then the operation of this Agreement and all obligations thereunder shall be suspended until such strike or industrial dispute is ended and the Company is able to undertake such inspections.

3. In consideration of the aforesaid services by the Company the Client/s will duly and punctually pay to the Company the annual sum of .............................. by equal quarterly payments in advance on the first days of January April July and October in every year.

4. IT IS MUTUALLY AGREED AND DECLARED AS FOLLOWS

(a) If any payment due to the Company from the Client/s shall not be paid within 28 days of becoming due the Company may without prejudice to the liability of the Client/s under this Agreement forthwith determine this Agreement and/or suspend until such time as the payments are made any further visits due to be made by the Company under Clause 1 hereof.

(b) This Agreement shall unless sooner determined by the Company as hereinbefore provided continue in force until determined by either party giving to the other three months notice in writing.

(c) If the Company shall at the request of the Client/s carry out any work or inspections or supply any materials other than those which the Company has agreed in Clause 1 hereof to carry out or supply the Client/s will in addition to the annual payments due under this Agreement pay the Company for any extra work carried out or materials supplied.

(d) The liability of the Company shall not extend beyond a duty to inspect and report any patent defects in the said plant and the Company shall be under no liability whatsoever in respect of any defects not apparent upon external examination and this Agreement shall not constitute a delegation of any duty of the Client/s upon the Company nor relieve the Client/s from any liability whatsoever.

(e) If at any time after the date of this Agreement and during its continuation there is any increase in the rate of wages, outworkers, allowances or other emoluments paid by the Company to its maintenance mechanics then the amount of the payments to be made under Clause 3 shall be increased by the same percentage that wages outworkers' allowances or other emoluments have been increased.

(f) This Agreement shall not be considered as an indemnity against accident and/or breakdown.

| SCHEDULE OF PLANT REFERRED TO IN THE AGREEMENT |
|---|
| |

AS WITNESS the hands of the parties this .............................. day of ........................
one thousand nine hundred and ..................

| | | | |
|---|---|---|---|
| For the Company | ........... | | ........... |
| For the Client/s | ........... | $2\frac{1}{2}$p stamp | ........... |

ENDORSEMENTS

# Index

*Italicised figures refer to related illustrations*